The Cyber Circuit

by James Bore MSc CSyP

A Collection of Articles on Cyber and Information Security
from Circuit Magazine

First printing 2024

1 3 5 7 9 10 8 6 4 2

ISBN: 978-1-7385153-0-1 (Hardback)

Security Blend Books

https://securityblendbooks.com

To my partner, Nikki,
tamer of unnecessary commas.

"If you think technology can solve your security problems, then you don't understand the problems and you don't understand the technology."

— Bruce Schneier

Contents

Acknowledgements

This is the part where I get to pretend I'm an award winner and ramble on for hours. If you always slip out to the toilet during the speeches, now's the time to turn the page.

First of all huge thanks to Jon Moss of Circuit Magazine, and Phelim Rowe, without whom I'd never have started writing these articles back in 2019 – and certainly wouldn't have kept them coming.

Thanks to my parents, for their support and encouragement, and the suggestion of collecting the articles into a book.

Thanks to my friends and peers who enthusiastically proofread and critiqued the draft versions of this book as I went back and forth on whether it was ready.

And finally thanks to Nikki, my partner, for (nearly) always respecting the office's 'busy' sign, keeping the cats out, and bringing me coffee.

Foreword

It never ceases to amaze me how we have reached the year 2024 and yet many individuals and organizations of all sizes don't think they need to take the growing cyber security threat seriously. "We're too small, so cyber security is something we don't need to think about, we're not going to get hacked," is something the owners of many small businesses still say today.

This is, quite simply, not true. Cyber security, or at the very least having a good cyber posture in any organization, is crucial to guarding against a potential cyber-attack. It isn't until an organization has a cyber incident or a data breach of some kind that they finally start to take cyber security seriously, but by this point it is too late. The damage has been done, not just from a financial perspective, but from a reputational one, which is often much harder to overcome.

One of the biggest things to explode into the mainstream in the last few years is the rise of artificial intelligence (AI), including ChatGPT and learning language models. Such is the power of ChatGPT that cyber-criminals are now using this tool to craft phishing emails, which makes it that slightly bit harder to spot if an email you receive is genuine or not. When we are all busy and rushing around it will be all too easy to click on a link that you shouldn't click on, especially if you are expecting a parcel from, say, DHL as an example and it reads well with good spelling and grammar. The growing cyber threat is something that no-one can afford to ignore any longer.

What struck me the most about this collection of articles that were originally published in "Circuit Magazine" by James Bore in 2019 is that while some things in cyber security have improved, and much has changed over the last five years, many things remain unchanged. There are still some myths that exist today in cyber security, such as:

You need a degree to get into cyber security.
> This is not true, as James outlines in this book. While a degree can be beneficial in some areas, it is not a prerequisite to getting into the cyber security industry as a career choice.

Strong passwords will keep my business safe online.

But what about multi-factor authentication? Have you implemented this in your business and for all your personal online accounts?

Implementing robust cyber security measures will cost too much.

This is also not true. There are many measures you can take to boost your organization's cyber security posture that cost little, and some are even free to implement.

Cyber threats are always external.

But what about insider threats? Have you considered your staff and/or suppliers, and how they might contribute to a cyber-attack or breach, even if invertedly?

"The Cyber Circuit" takes these myths, breaks them down and presents them in an easy-to-understand format that cuts out the complexities and dispels misconceptions in cyber security. It can also help to empower individuals and organizations of all sizes to better understand, implement and adopt more effective cyber security measures. Demystifying cyber security today is a collaborative effort that involves clear language and communication, community engagement, transparency, user-friendly tools and cyber security awareness and education. Only then can we as an industry go some small way to thwarting the cyber-criminals.

- Lisa Ventura MBE Founder – Cyber Security Unity

Preface

In 2019 at a security convergence event (one of those ones where we talk a lot about how security is security, whether it's cyber, information, physical, biological, chemical, environmental, or whatever else you choose to apply it to) I was introduced to Jon Moss.

At this point I was pretty comfortable with where I was in my career. It would be a while longer before I hit my lowest and most stressful four months, which would be my worst experience of employment and the kick that finally made me look for other ways to work. That's another book in itself.

It was also shortly after I completed my Masters — and the story of why might tell you something useful about the viewpoint these articles are written from. Originally I dropped out of university way back in the 00's, and went into work (network manager for a school, having their data protection issues and ISO 27001 compliance dropped on my lap immediately was very much a case of jumping in the deep end).

Over the years my lack of a degree never caused any issues, and then around the mid-10's I was interviewing someone for a role.

They were a new graduate, and confidently and authoritatively told me that no one could or should work in information security without a degree.

As a note for anyone looking to get a job, this is not a good tactic to take – at the very least without knowing whether the person interviewing you and making a decision on whether you get the job has a degree.

That incident stuck with me, and a few years later when the money was available I decided to enrol on a part time Master's course with Northumbria (another tip, if you're looking for them, most universities will take professional experience as a substitute for a Bachelor's degree). While I came to enjoy the degree, the motivation for starting it was very much spite.

I'd also been dabbling in writing for years, from blogs to short stories to essays, and played with the idea of going freelance with it. This was one of those moments everything came together as we started talking about the British Bodyguard Association, and this Circuit Magazine which Jon published for close protection professionals. Jon asked me if I'd be willing to write a few articles to introduce people in the physical domain to the one I worked in.

Frankly my concern at that point was more about how dull my field would seem in comparison. I think that's a problem many people deal with, and the years since have convinced me that people outside of cyber and information security find it much more interesting than I ever expected.

Either way, I agreed, and at the end of the first year just... kept on writing them (sometimes, okay often, with a bit of prompting to remember to send them in time). It meant I got to talk about a number of pet topics that I just thought would be interesting, and discuss some of the problems and issues in the cyber security industry with a receptive audience.

And here we are – five years later and I realise there's been enough writing to turn it into a book. Some of the articles have been updated, some have been extended, there's a handful of extra ones in here (yes, exclusive bonus material) that never got finished because cats distracted me, or they never got off the whiteboard.

In cyber and information security a lot can change over five years, so where I've updated articles I've tried to keep the original spirit as much as possible. Where my own views, thoughts, or knowledge have evolved since then I've called out that change to be as transparent as possible (and because I think the changes in the way we look at things are interesting).

So here it is, five years of ramblings and thoughts on cyber security and the bits of it which I think are important for people to know to protect themselves. I hope you at least enjoy it, and at best find something of value.

— James Bore MSc CSyP

Introduction

The first cyber attack occurred in 1834. Criminals stole financial market information by accessing the French telegraph system.

The first report which stated the problem clearly was the Ware report[1] (more accurately, and less interestingly, Security Controls for Computer Systems), in 1970.

It would take two more years for the Anderson report[2] (again, less interestingly titled Computer Security Technology Planning (Volume II)) to start talking about approaches to dealing with the problem in 1972.

And then we have 1988, often chosen (wrongly, at least in my view) as the start of cyber security, when the Morris Worm was released. The worm was meant to be (supposedly) a harmless experiment, and within 24 hours had impacted a whopping 10% of the internet. That sounds less impressive when you consider it was at the time only 6,000 of the 60,000 connected computers, but it's still significant.

---

[1] Ware, "Security Controls for Computer Systems (U): Report of Defense Science Board Task Force on Computer Security"

[2] Anderson, "Computer Security Technology Planning Study (Volume II)"

Since that first attack the evolution of our technology has gone hand in hand with the evolution of cyber crime, though it really got a foothold in the last century with the spread of the computer and the development of the internet.

At its heart, cyber security is no different from any other form of security. All security comes down to people, and the technology is just a set of tools to connect people to other people. The ease of that connection is what makes cyber crime such an effective industry, and make no mistake it is an industry – currently estimated to be worth around $10 trillion annually (although when you get to the article on FUD you may have some questions about how it's calculated and whether it means anything).

It's also important to know that much of the cyber security industry is not in competition with cyber crime, no matter how much it may seem that way. Instead cyber crime is what creates the market for cyber security vendors, along with technology not having been designed and built with security in mind. Overall, people are left under threat by decisions completely outside their control.

The following articles will cover some of these issues, along with talking about current, past, and future trends in cyber and information security. They are written to provide an introduction to numerous diverse areas, and while not intended to provide a perfect understanding they are written to present the issues in an understandable manner rather than obscuring with technical jargon.

I considered various ways to present them, from grouping them by theme to alphabetical, and ended up going with chronological order for simplicity, separated by year. The Cyber Circuit has not been written or collected with the intent of being read cover to cover, more to dive in and read whatever topics takes your fancy or is relevant at a particular time. Each article will provide you with enough information to get started, and find out more.

More importantly, reading these will give you a better understanding of the reality of the world we live in, and the ability to see through the science fiction hype that is misleading. Conspiracy theories, disinformation, and anxiety are often a result of not having clear information and taking the time to understand will enhance your learning and identification of tools and techniques to protect yourself and others.

The chronology of the articles does tell a story. Early articles consider some of the broadly applicable fundamentals, particularly ones that could affect people using close protection details. This went through a general introduction, social engineering, the idea of Man-in-the-Middle (MitM) attacks, smart housing troubles and social media, capping the first ones with a brief introduction to the idea of threat modelling.

I was also relatively early in my professional writing career, and especially in writing for The Circuit Magazine, so I was finding my way. You may notice the style of writing evolves and relaxes, and even the topic selection as I gain more confidence with the audience over the years.

In an earlier draft the articles were vaguely grouped into topics, which never felt quite right. They're now in chronological order and grouped by year, and you can find the titles in the table of contents. If you're looking to see where a particular topic is discussed there's an index in the back.

The Unseen Articles are various articles personal to me and exclusive to this book. Some were originally started but left unfinished when I came across a new idea. Others never made it off the whiteboard. A few were finished, before I decided they didn't quite fit with the magazine, or the topic was less relevant. One or two might be based entirely on a pet peeve that triggered me to write something for my own satisfaction, but didn't go any further.

■ You'll also come across these, which are something like DVD commentary. If you see a box like this then it's me adding commentary which wasn't part of the original article, but which I thought was important, useful, or felt like adding.

In the Appendices section you'll find a few extra bits and pieces. A collection of book recommendations for further reading, some useful resources such as tools or learning platforms to explore more about cyber security, an index, a bibliography of references, and a list of all the figures in this book, and of course a glossary to deal with the many, many jargon words and acronyms in security where I've used them.

And of course, if you've gone for the electronic version rather than print you'll have clickable hyperlinks. Beta-readers have told me that if you click these, try to remember the location you came from, as sometimes the e-reader back button doesn't work and it's possible to get lost.

Welcome to the anthology, and enjoy the articles.

Part I

2019

Understanding Cyber Security

Issue 46: 2019-02-21

Welcome to the first in a series of articles looking at the world of cyber security and how it can both benefit and harm efforts in security and protection generally.

The electronic and physical worlds are converging ever faster, with smartphones, cars, cameras, aircraft, drones, and even houses relying more on information technology to make lives more comfortable. The field of information or computer security has been around for a while. As this crosses over into the physical world, we've now rebranded it as cybersecurity – bringing in cyber-physical systems such as industrial control systems.

Cyber security is often seen as a niche area that requires a lot of specialist knowledge to apply. This is partly true – in order to configure a web application firewall, someone needs to understand how to work with the technology at a very low level. What is often missed, as the technologists take over, is that cyber **is** still security, and the same fundamental principles apply to designing and building effective protections.

The basic principles of cyber are simple and can be understood without a deep dive into the huge range of technological applications that exist today.

1.1 Risk Management, not Prevention

It is a truism in cyber security that you cannot have perfect security. Any system is vulnerable, and the goal is to make carrying out an attack cost more than the expected benefit. At the most basic level, it is risk management and mitigation rather than attempting to eliminate risks entirely. Different technology solutions exist to reduce the impacts of different threats, and a lot can be done simply by ensuring processes and systems follow good principles.

> ■ Originally I didn't delve into the good principles here, but they're straightforward enough. Ultimately, harmful security events are the result of unanticipated, undesirable behaviour. When building systems many models only consider whether they fulfil their requirements, to render them secure it's a matter of making sure they fulfil only their requirements.

1.2 The Weakest Link

Given it is about risk management, we have to focus prevention on the weakest link in any system. Spending millions on a top-of-the-range firewall with real-time monitoring and a follow-the-sun operations team to protect a piece of data is only useful if no one is printing out copies of the data and throwing them into the dumpster around the back of the office. Dumpster diving is a time-honoured tradition among attackers.

1.3   Human Vulnerability

The hardest weakness to address is simple human fallibility. Training and awareness in how to take basic precautions against attacks are essential to preventing them – the technology can never be perfect. While in the industry, we don't expect humans to be perfect and can put some technology in place to help with this, if they aren't given the information they need to protect themselves, then we may as well throw out all our expensive toys and go home.

> ■ Over the years since this article I've taken a serious dislike to describing humans as the weakest link in any way. It's still common throughout the industry, and you will hear people report 'humans are the weakest link' far more than they should.

1.4   Least Privilege, Minimum Access

One advantage of working with technology is that trust is absolute – you either have it or you don't. The least privilege principle (along with its other names) is theoretically simple[1] – you only get access to information, systems, areas, or anything else when it is absolutely essential to carry out a role you are trusted to perform.

This applies to people accessing systems just as much as inter-system communications, and any system implemented with a well-considered least privilege model during the design stage will be an order of magnitude more secure than an open system with all the firewalls in the world.

1.5 Incident Management and Response

Being blunt, in cyber security, we have to accept that we will fail, repeatedly, and will fall to attacks. We face an asymmetric threat as any person or organization, the moment that an organized attacker turns their focus on us. This is why, once the basics are in place, being able to detect and respond to an incident is key. The average dwell time of an attacker on a network, with full access, is somewhere between 50 and 150 days, and some attacks have gone undetected for multiple years.

> ■ Dwell times are now better understood, and vary between different attacks. Ransomware incidents often have dwell times less than 24 hours, while more sophisticated attacks can dwell undetected for years (and sometimes years more after detection)[2].

---

[1] But more complicated in practice as it needs a full understanding of system functions.

Most of the technical expertise in cyber security is about know-ing how to, or finding ways to, apply these principles. If this is done early in the design stage of a system, the need to layer expensive security solutions on top of it later on when it gets breached is massively reduced. The same applies to implement-ing processes and procedures.

One final important note is that cyber security is not the same thing as information security. Information security, or infosec, is concerned only with protecting confidentiality, integrity, and availability of information. Cyber security overlaps with infosec but extends into areas where systems and technology interact directly with the physical world, and damage may be dramati-cally more extensive than lost information, while not touching on areas which do not involve technology[3].

1.6   Examples

Stuxnet[4] is the most advanced malware discovered to date[5] and is believed to have damaged a fifth of Iran's nuclear centrifuges over the course of two years before it was discovered – it was carried into the facility on a USB drive.

In 2014, unknown attackers caused serious damage to a German steel mill by preventing the blast furnace from being shut down

---

[2] Mandiant, *M-Trends 2022*

[3] At least that's the theory – in practice the overlap is so large, and the definitions are so poorly understood by most people, that they border on interchangeable terms.

[4] Kushner, "The Real Story of Stuxnet"

[5] This is still true over a decade later.

as scheduled[6].

In 2018, the White House publicly acknowledged that Russia had infiltrated, and potentially had control of, some utility control systems including power[7].

Today we get more and more warnings in the news of home automation systems, including surveillance systems, being compromised and accessed by attackers for malice or mischief. Modern vehicles have a lot of automated intelligence, and demonstrations have been carried out to switch off braking systems, control acceleration, and take over windscreen wipers – self-driving cars will only add to the possible attacks. Malware on mobile phones can discreetly turn on cameras and microphones with one unlucky download, turning a personal device into a surveillance device[8].

All of this is possible because security was not considered a priority by device manufacturers, whether of cars, home automation, phones, or even medical equipment. Knowing the potential for attack through these vectors is vital for personal and corporate security to weigh up the risks and take action to bring them down to an acceptable level, or prepare for the possibility of an attack.

---

[6] "Hack Attack Causes 'massive Damage' at Steel Works"

[7] *Russian Hackers Infiltrated Utility Control Rooms, DHS Says*

[8] This is possible, but far from as simple as fiction would lead people to believe – and a common issue now is people convincing themselves that it has been done to them.

> ■ In five years of writing this column, this has not changed as much as anyone might hope. If anything, it's worsening.

In the next article, I'll be diving into an area of cybersecurity that is less technological, and more about the weakest link in any modern system, with an overview of how attackers use social engineering to shut down accounts, discover personal details, and generally cause havoc.

Social Engineering and Cyber Security

Issue 47: 2019-05-09

Welcome to the second article in this series looking at cyber-security and its interaction with the protective and wider security world. This time we'll be looking at a collection of tactics used by attackers to bypass security technology by targeting the weakest link — the human in the chain.

Social engineering is essentially a technical term for knowing how to persuade people to do something against their best interests or better judgment. For example, talking someone into handing over the password to their email account whether in person, via email, text messaging, phone call, or simply watching over their shoulder as they type it out[1]. Another example might be persuading a mobile provider's service centre to shut down someone's account through impersonation.

Broadly, there are only a few methods in social engineering,

---

[1] Respectively, because we always need more jargon, phishing, SMishing, vishing, shoulder surfing.

and different ways to apply them. The most commonly seen these days, due to its anonymity, effectiveness, and the ability to automate the process, is various forms of phishing. Just for some brief history, as I get asked this, the misspelling of phishing to mean this type of attack comes from the 1990s and followed the use of phreak to refer to people who exploited phone systems.

2.1 Phishing, Vishing, SMishing, Spear Phishing, Whaling

The only differences between the various forms of phishing are the targeting and the method. Phishing, spear phishing, and whaling all rely generally on emails, whether they're spoofing, stealing genuine email addresses, or simply casting a wide net and using a disposable email account. Spear phishing (Figure 2.1) and whaling take a little more effort and often involve researching the target.

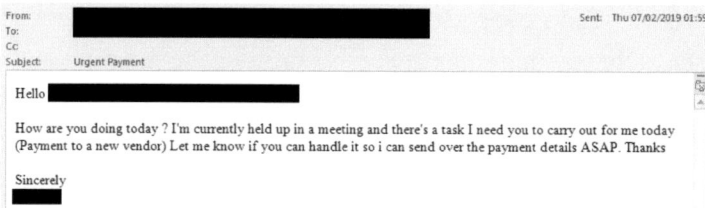

Figure 2.1: An example of a spear-phishing email. Details are hidden to protect the target. The attacker used an email domain with one character difference from the genuine one and impersonated the CFO to an employee in finance. Image credit to the author.

Generic phishing and SMishing tend to cast a wider net, relying on large numbers to catch a victim to exploit, while vishing uses a phone call. All of them rely on a tactic referred to as pretexting or impersonation to persuade people to do what the attacker wants.

Watering hole attacks are another variant of phishing where instead of asking a victim to disclose information or to download malware, an attacker will ask them to visit a website. This may end with a compromised machine, a compromised password, or a compromised user, but never well. Whaling is similar to other phishing tactics but deserves separate mention for the sophistication of the attacks. Those who whale, targeting high-profile executives, are organized and will do significant research and preparation, using open-source intelligence such as LinkedIn profiles and company filings to construct organizational charts of a company's internals. Reconnaissance will be performed by vishing at low levels to understand how the company works and possibly to gain access before an attack is executed. One particularly well-executed attack involved company registrations to match existing suppliers and is known to have cost several top technology firms at least £77 million[2].

For a good example of some of the most common tactics, Google has provided a quick quiz at https://phishingquiz.withgoogle.com/.

---

[2] At the time of publication the court case hadn't completed. Now it has, it's known that at least this amount was lost to the fraudsters by Facebook and Google. Most frauds do not get prosecuted so quickly, but when you're a large enough company things tend to move a little more than for the average person.

> ▰ Catfishing has fallen out of use recently, much to my relief, and the much more useful and easily understandable term romance fraud seems to have taken over.

Phishing is not just used for quick financial gain but has been used for particularly vicious blackmail campaigns – usually in a form called catfishing. In catfishing, an attacker will create a profile on a dating site designed to be appealing to a particular group of people. Sometimes it will be entirely false, others true enough to allow them to arrange video calls. Once a relationship has developed, the catfish may go either for a scam, saying they need money for a plane ticket, for example, or may descend into particularly vicious blackmail using previous intimate messages, images, and video exchanged. Often these scams are not detected, as victims are too ashamed to report them.

I have no idea why catfishing isn't referred to as catphishing.

2.2 Pretexting, Tailgating, Baiting, and Quid Pro Quo

While phishing does make use of pretexting, it is generally given its own category in any attack frameworks. Phishing is electronic and, in a way, simpler. Pretexting might be done via the phone, where it crosses over with vishing, or in person. There are long texts written on how to pretext, getting uniforms from different commercial companies, how to walk right to avoid too much attention, hi-vis jackets, clipboards, the right sort of conversation to make to be forgettable, and a lot of work and theory on influence and persuasion.

Tailgating sometimes falls under pretexting, with an attacker simply following people through what should be a secure door – often by carrying a 'heavy' box, rushing for the door, or spending ten minutes in the smoking area with a group and a fake badge.

Baiting is usually used to gain greater access or information once an attacker has performed some basic compromise. The classic example here is dressing as 'the IT guy' for a large office, yanking the network cable for someone's machine, waiting for them to ask for help, and then simply suggesting they get a coffee as this will take a while. It is a method of setting up an opportunity to further compromise a target, and closely related to quid pro quo.

Quid pro quo can follow a baiting attack or be entirely separate. It is a method that works on the basic human need to reciprocate help that we receive – whether that's fixing a computer or being given a gift.

Some penetration testing companies will offer physical pen testing and red teaming, where they will try to exploit all of these tactics (and more) to gain access to a designated target before reporting exactly how they have done so and where to improve in the future.

I have very rarely heard of them failing to get access to a target area, and they are always pleased (and surprised) to have an attack effectively shut down before it completes.

> ■ Despite the prestige of physical penetration testing, in recent years questions have been asked about its value outside of high value targets. This is because it's an expensive activity and attackers willing to try physical penetration tactics against your average organisation are few and far between.

Whole books have been written on social engineering tactics, some of them well worth reading. At this point I'm only trying to give a basic overview so you will be better informed of some of the methods and how these relate to both cybersecurity and wider security. Next time I'll be digging into man-in-the-middle attacks, a less common but devastatingly effective method of compromising information and people.

The Man in the Middle

Issue 48: 2019-07-08

Welcome to the third article in the series looking at introductions to cybersecurity. In this article, we'll explore a type of attack most people will be familiar with in principle, if not in technical practice: the MitM. The basic idea is simple – an attacker sits between two trusting parties, intercepting their communication and impersonating each to the other. While this may be somewhat harder in face-to-face situations, even a phone call provides potential for eavesdropping or impersonation.

Becoming the Man in the Middle is more challenging with some technologies than others and hinges on somehow being positioned in the middle of the connection. With computers communicating over a network, a technique called Address Resolution Protocol spoofing makes this relatively easy. In this method, the two sides of the conversation are fooled into sending their messages to the attacker's computer rather than each other. Even when encryption is used, each computer believes they are talking to a trusted recipient, assuming all passwords and keys are trusted.

The restrictions are that the attacker must somehow have access to the network between two people and be able to successfully impersonate them. Physical security methods can help prevent this, especially in scenarios where access to an office is required. Unfortunately, with the increasing reliance on wireless networks, an attacker may only need a malicious node within range. Even in cases where wireless networking isn't used, the tools needed for a MitM attack are easily and cheaply available.

3.1   Tools

Figure 3.1: The WiFi Pineapple is a popular tool for wireless networks. Image credit to Hak5

The WiFi Pineapple (Figure 3.1) is probably the most famous of these tools, particularly after its alleged use by GRU intelligence units to break into the networks of the World Anti Doping Agency, a nuclear energy company in Pennsylvania, and the Organisation for the Prohibition of Chemical Weapons[1]. The cheaper, smaller WiFi Pineapple shown here can be acquired for around \$100. With a mobile phone and an appropriate USB cable, it's possible to sit in a public place and intercept the network traffic of everyone around you without anyone noticing.

■ While the tools for these attacks are cheap and easily available, the motivation to carry them out is much less common. Your odds of encountering this attack in real life, unless you have reason to be targeted by someone technically capable, are almost non-existent. Even then, modern security measures lessen the impact of these attacks through various means, making it harder for an attacker to simply snoop on your network traffic without additional steps.

In most cases an attacker trying an MitM attack needs to be physically present, meaning they are much riskier for the attacker than almost any other method.

---

[1] Burgess, Matt, *Russian Spies: How Russia's Top Secret Global Hacking Operation Unravelled* | *WIRED UK*

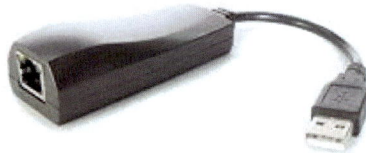

Figure 3.2: The LAN Turtle is a popular tool for cabled networks. Image credit to Hak5

Another popular tool, more suitable for cabled networks, is the LAN Turtle (Figure 3.2). Essentially, it's plugged into the back of a computer's USB port, a network cable is plugged in, and unless it's discovered, an attacker then has their own hostile computer on the network – almost invisibly unless detection keys are used. The LAN Turtle is available on Amazon for about $50.

Figure 3.3: A Stingray device is effective for mobile phone networks. Image credit to Simone Margaritelli.

Of course, neither of these tools will help with mobile phone networks. For that, a Stingray device, once a top-secret tool, is more effective. A professional Stingray (Figure 3.3) device comes with various restrictions and high pricing. Building your own involves some knowledge, a laptop, about $20 of parts (available on Amazon), and about half an hour of time.

3.2   Finding and Beating the Man in the Middle

There are ways to beat the Man in the Middle. Website certificates are becoming more and more common every day and go a long way towards at least warning users. However, it's common for people to simply click through and ignore the security warnings now built into most browsers. Since the attacker can put up their own false certificates or simply strip certificates out of the equation entirely, training on how website certificates work, what to look for, and what errors mean is crucial. In theory, a certificate is only issued to people who can prove that they own the website it's used for, and mostly this theory holds true.

Figure 3.4 and Figure 3.5 show how these can appear in browsers, and the details to look at.

> ■ Various campaigns such as HTTPS Everywhere and Let's Encrypt[2] have changed the picture on this, and warnings from browsers about accepting untrusted certificates have become much more visible. Having a certificate is no guarantee of security, since they can be issued to malicious sites, but at least you will know you are being securely defrauded if you visit one.

---

[2]   Aas, Josh, *Let's Encrypt*

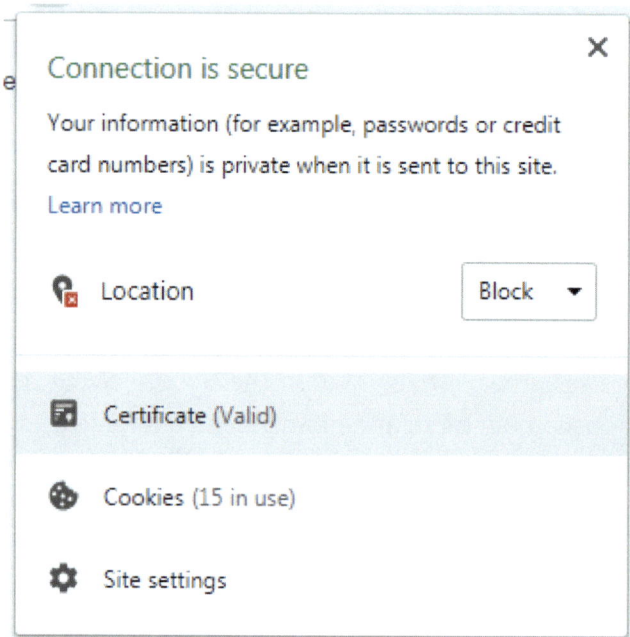

Figure 3.4: A first check is to make sure the certificate is valid – normally you will be warned if it isn't. Image credit to the author.

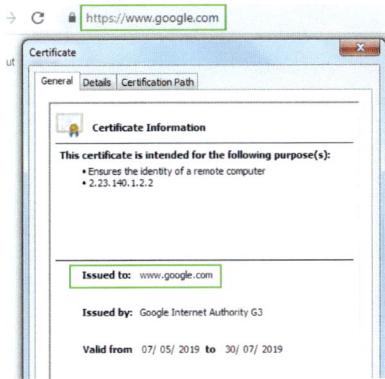

Figure 3.5: You can see that the certificate was issued to the site it's being used for – it's now down to whether or not you trust that the issuer checked who they were granting it to, and that no one has managed to steal their signing certificate. Image credit to the author.

Certificates aren't as helpful if you're concerned about a phone call or text message being intercepted. Luckily, there are many solutions to provide encrypted calls, chats, and text messages, varying in price and trustworthiness. WhatsApp is one of the more popular ones, though there are some serious security concerns being raised around it.

My personal preference, both for price (free[3]) and effectiveness, is a system called Signal. Signal works on Android and iPhone as well as desktop, provides end-to-end encryption for text messaging, and covers phone calls. One important feature provided is the so-called 'Safety Numbers', essentially a password you can exchange in person or through some other mechanism to confirm that the phone at the other end is the one you're expecting. Anyone attempting to hijack the communication after a num-

---

[3] You can now donate to get a few extra features, but the basic app is free.

ber has been verified will alert you that the number is being changed. Of course, many other options are also available.

Insecure Smart Houses

Issue 49: 2019-08-31

For the fourth article in the series, we're going to be looking more forwards at some emerging threats out there. They are only of limited relevance today, but as the technologies involved become more widespread and implemented into every facet of life, they will only become more prevalent. While it sounds like the stuff of science fiction, these threats exist now and are not going to go away.

> ■ Despite my alarmist comments here, and the fact these threats are still theoretically possible, the biggest threats to your smart home involve someone accidentally stealing a lightswitch, or deliberately stealing your smart doorbell.

For simplicity, we'll say that a 'smart' device is anything which connects to the internet (or a network) and is not intended to be a computer interface. Intended is the key word there, as many of these devices are insecure for the simple reason that they are a computer. The problem is that it is now cheaper and easier to put a general-purpose computer into a device and run some software to, for example, turn lights on and off than it is to design a single-purpose lightbulb that also connects to a network.

The side effect of this is that the smart lightbulb connected to the network is also a computer, and one that the owner has very limited control over. Every 'smart' system connected to a network increases the potential attack surface and potential vulnerabilities. A lot of the time this is overlooked with the argument that, for example, a lightbulb simply isn't a huge security threat – even if compromised, the worst someone might manage would be to turn the lights on and off. Unfortunately, things are never that simple.

Without even looking at risks linked to specific devices (we'll get to those shortly), many smart devices rely on connecting to a wireless network and will happily connect to a spoofed network while revealing the password to all and sundry. We saw in the last article how a MitM attack can be used to this end, and a smart device is no less vulnerable than any other.

Last time we covered MitM, but there are much more interesting and dangerous issues with many of these 'smart' devices. We'll look only at those which could arguably be called home devices – semi-autonomous vehicles, drones, and similar could easily fill a series on their own. For brevity, we'll look at three basic threats which come up with 'smart' devices – their use for reconnaissance, the potential for mischief, and the potential for physical threats to security. I separate mischief and physical threats only because the people likely to exploit these are two separate groups; pranks and mischief are much more likely for opportunistic attackers with no real targeted motivation, while the dangers of physical damage are much more likely to be a sophisticated, targeted attack of types we are not really seeing on a large scale as yet.

> ■ Again, with the way these threats have actually developed I now cringe at how much hype I added here. Yes, these threats exist. Yes, they're fascinating, exciting, and sexy.
>
> No, realistically you're not ever going to see them – the worst I expect to see in coming years is ransomware attacks against appliances. Imagine having to pay out bitcoin to get your dishwasher working again. For industrial control systems and office smart devices there's a slightly more realistic threat.

I occasionally do presentations on the dangers of 'smart' devices, or the Internet of Things (IoT), and usually start with a list of the absurd devices that can now be connected to a network. It's a good place to begin, so let's imagine a home equipped with all of the latest devices, from intelligent electronic locking systems with video doorbells through to the 'smart' egg rack in the fridge. No, the egg rack example is not a joke, though I will admit I've yet to work out a potential threat from someone managing to compromise one.

Our imaginary smart home will be equipped with the following:

- Doorbell with the ability to start a video call to the owner's mobile when pressed

- External CCTV (basic commercial system, wireless cameras with an online portal to be viewed remotely)

- Electronic door lock, able to be unlocked remotely by mobile

- Smart lighting everywhere, remotely controlled and scheduled

- Automatic pet feeder/treat dispenser with video camera and screen (to check on and talk to the pet when not home, and dispense treats on command)

- Aquarium with mobile-controlled pump and thermostat

- Video-enabled toothbrush, for more effective brushing (this really does exist)

- Smart kitchen appliances, including a fridge, kettle, toaster, oven, microwave, dishwasher, tumble drier, and washing machine

I've left out quite a few potential devices since they'll either be of limited use to an attacker, other than the MitM potential mentioned earlier, or duplicate some of the above.

So first of all, let's start with everything with a video camera. Whether through an MitM attack or direct exploitation, most cameras which have not been set up professionally are highly insecure (and a number of those set up by professionals who have not been trained in how to configure IP cameras, since the default settings leave a lot to be desired). Those which are built into smart devices, such as the toothbrush and pet feeder, potentially even more so.

The good news is that the toothbrush relies on Bluetooth rather than wireless networking, so an attacker looking to make use of the spy camera in your bathroom would have to be within range, or have a booster within range. That means they're limited to being within a few tens of meters, or having a small battery-powered device within that range. Of course, our hypothetical attacker would also need to find a suitable exploit to hijack the toothbrush's camera, but given that manufacturers of these devices are not known for their diligence in designing secure systems, and I have yet to hear of someone connecting their toothbrush to the internet to patch it, someone with the appropriate motivation is likely more than capable of doing so.

The pet feeder, while at least in a less sensitive area, is more of an issue. Since it is designed to connect to a wireless network and allow the owner to both see and communicate through it (of course, having a microphone as well so that the pet in question can speak back) using their mobile phone while away from home, the only thing that prevents an attacker from doing so is the authentication. Some devices or services are so poorly designed as to allow access without any authentication, or use a universal default username and password to grant access to the vendor's backdoors in the system (which are at least used to update the software).

> ■ It was a few years after writing this that Amazon AWS suffered an outage[1]which disabled our automated cat feeders, and so now I have a new threat to add. It is being woken up by angry, starving cats because a data centre failure halfway around the world has managed to prevent them getting their last two meals. That is what happens with poor design.
>
> And to add to the misery, if the company stops supporting the device, or shuts down for any reason, your smart device is likely dead with it.

The CCTV cameras, while also having many of the same issues, highlight another problem with these smart devices. The search engine Shodan (https://shodan.io) allows users to find open devices connected to the internet by type (including automatic number plate recognition systems, wind turbines, and CCTV monitoring nuclear power stations), and where weak devices are used provides a very convenient list for those looking to cause damage. Denial of service attacks in recent years have moved away from making use of individual's computers, and some of the largest attacks have been enabled by running a botnet consisting of poorly secured CCTV cameras, usually without any monitoring by the owner. This has led to some of the largest attacks ever seen, and they continue to escalate[2].

The video doorbell, of course, causes the same sorts of issues as any other camera-enabled system but has a whole new set of risks associated when it is connected to a smart locking system.

---

[1] Moss, *AWS US-East-1 Lamda Outage Causes Issues Globally*
[2] Cloudflare, *What Is the Mirai Botnet?*

Not only are these systems easy to hijack physically (earlier this year I gave a presentation on how to 'pick' electronic locks based on RFID cards), when they are connected to a cloud service which allows unlocking through a mobile phone (and sometimes through Bluetooth) the problems become obvious.

At this point, a dedicated, sophisticated attacker essentially has video surveillance and audio monitoring throughout the house, yet we aren't quite done as there are some more esoteric attacks against smart devices which can dwarf the simple risk of espionage. The kettle is a prime example of this – initially it does not seem a particularly dangerous device, at worst someone could maybe cause a cold cup of tea or keep re-boiling the water. It's the second one that becomes a problem (not with all smart kettles, but there are models which have this combination of flaws) where the thermoregulator is implemented in software rather than hardware and runs on the same general-purpose computing as the 'smart' controls. Someone wanting to cause real damage, aware of these issues and how to exploit them, could turn off the safety cut out and repeatedly boil the kettle until it ran dry. At that point, continuing to heat causes a clear fire risk[3].

The same sort of risks apply with the toaster, oven, and some other smart appliances. Another example is, say, a smart freezer – a subtle attacker could cause it to thaw and refreeze overnight, repeatedly, spoiling any food without the owner's knowledge. While food poisoning is not as immediately dangerous as a kettle bursting into flames, the risk is clear.

---

[3] Marks, Paul, *Killer Kettles Show Security an Afterthought for Connected Homes* | *New Scientist*

Unfortunately, at the moment there is no real remedy to many of these problems, except a comprehensive security review of any 'smart' devices being purchased by experts. Such a review gets expensive quickly, though many pen testing companies showcase their capabilities by highlighting the risks in commercial products (and in 'smart' toys, such as teaching a doll to swear or a teddy bear to share video with strangers) and making them publicly available. Another possibility to reduce the risk would involve making the house network highly secure and isolated, with only a well-protected connection from the phone, but as many of these devices rely on cloud services for management this would remove any benefits of the intelligence.

> ■ New laws are being passed to improve this scenario, but they are currently at the level of removing default passwords and ensuring that updates can be applied. It's a good step in the right direction, but still only a first step[4].

What is most needed is education about these risks and a demand from manufacturers for demonstrably secure products based on the best practices established over years of information and cybersecurity. Without this, we will continue to see the attack surface multiply year on year, as more and more 'smart' devices are turned against their owners. For the time being, knowing the risks these devices pose, being aware of how they can be misused, and knowing how to isolate them when necessary (or simply not purchasing them) will have to do.

---

[4] DCMS, *New Smart Devices Cyber Security Laws One Step Closer*

In the next issue, we'll be looking at a brief introduction to how you can hide (to a point) from Open Source Intelligence (OSINT) and make information on yourself harder to find. We won't be covering the full scope as the field is huge, just a few simple basics to start with and some recommendations on where to find a lot more information.

Hiding from OSINT

Issue 50: 2019-10-31

In terms of cybersecurity, OSINT covers any data or information that can be collected from publicly available sources. It often comes as a surprise just how much is available and the nefarious uses it can be put to. OSINT can be applied towards defensive purposes, but we will be looking only at malicious purposes. One of the biggest challenges of OSINT is not merely recognizing it as a threat, but encouraging the behavioural change needed to protect against it widely enough. It is not enough simply for a principal to stop posting Instagram pictures of their travels to hide them; their colleagues, friends, family, and employees also need to be aware of the need to take care with information that could be misused.

5.1 Social Media

The first and simplest step is to look at any social media sites in use and fully review any privacy settings available. Depending on the site and the network of connections, different settings may be appropriate. The important idea to remember is that only information that someone is happy to share publicly should be put on a site. Even where details are shared only with connections, friends, or family, the target of any OSINT operation is then relying on the security of their connections to protect their own information.

Sharing pictures of family holidays is a common activity on various social media platforms, and when combined with a home address or check-ins at locations near to home, this can inform a malicious party of a valuable target property that is currently unoccupied. Burglaries are not the only options, as an unoccupied property is also useful for people looking to protect themselves while committing various forms of fraud by having valuable deliveries sent to an address they are not linked to. Photos and videos of Christmas present openings will be common in a short time window, and unwisely shared are very popular with thieves with shopping lists[1].

Even when not providing targets to a potential burglar, sharing personal data can be a serious issue. When phoning a bank or speaking with a phone company, often personal information is requested as a security check. Guidance for these security questions often suggests examples like the below:

---

[1] Homewatch Group, *How Do Burglars Use Social Media to Find Targets? - Homewatch Group*

- What is your mother's maiden name?

- Where was your first school?

- What is your birthdate?

- What was your first pet's name?

> ■ You would hope that in the years that have passed since I wrote this, the security question guidance would have improved. You would be hoping in vain. One option of course is making up fictitious answers, but if you do this and forget them you have a different problem.

Answers to all of these questions are easily available through social media postings, and it is important that a principal is aware of this either when setting up the security questions (in which case an inaccurate memorable answer can be provided) or when posting information.

One particularly helpful action if there is a good relationship with a bank or service provider is to request notification any time someone answers these questions inaccurately. Unfortunately, many do not offer this service.

5.2   People Searches

While social media is the most obvious and often the first target for OSINT, it is important to recognize that it is not the only source. Various people search engines, both legal and otherwise, compile various sources of public information such as electoral registers, company filings, news reports, and others and tie them to individual identities as much as possible. These are often commercial platforms that will charge a small fee for a search, but the available information is worthwhile. As an example, a search for me on one of these platforms reveals my name, address, house price, and positions as a director.

> ■ Since this was written searches on me will turn up a lot more. One of the unpublished articles included in this book talks about informational chaff (chapter 26) and how it can help if your concern is more frustrating OSINT than reputation management.

These details are pulled from the UK electoral register, Companies House filings, and property search sites. Each of these requires a different approach to prevent disclosing the information, and for many people, the effort involved is not worthwhile. When it is worthwhile, in many cases services have an option to opt out of publication. Where they do not, such as Companies House, the only way to hide some information is to have a separate registered business correspondence address.

There are other methods of authentication now, popularly sending a one-time password over SMS. As we'll see, this is far from a guarantee of safety and means that for someone truly trying to protect themselves against particular attacks, it is vital to have a secure phone number with no connection to the individual.

5.3   SIM Swapping

One of the rapidly growing attack methods is the SIM swap. While this goes beyond the scope of OSINT, it is only possible because attackers can put together information to enable the attack. At its simplest level, SIM swapping is an impersonation attack – either in person or by calling customer services for a mobile provider. Using publicly available information such as birth dates, an address, and a phone number, along with a few other pieces, the attacker persuades the mobile provider that they have lost their SIM card and need a new one. The moment they have that SIM card, they have access to the target's mobile number[2].

When SMS tokens (single-use passcodes via text message) are sent to provide 'secure' access to systems, they are sent to the active phone number. It's easy to see how a targeted SIM swap attack can grant access to vitally important systems. The best protection is simple – I have a dual SIM phone, with a second pay-as-you-go number on a separate provider that is used only for these services. Since there is nothing tying the number directly to me, it becomes much more challenging for an attacker to carry out a SIM swap.

Ideally, providers would start providing better protection against this attack vector by requiring stronger authentication and using different methods than SMS messages to access accounts, but

---

[2] Tims, "'Sim Swap' Gives Fraudsters Access-All-Areas via Your Mobile Phone"

until this happens[3], a separate unlinked phone number is the best method I have found.

5.4 Finding out more

While limiting easily available information and separating authentication phone numbers from known ones are two simple and effective tactics to prevent opportunists from using OSINT, when targeted by sophisticated professionals, things become more complex. Dealing with the capabilities of a well-motivated investigator is far beyond what I can go into in a short article, but there are very useful resources to look into for more information.

- *Hiding from the Internet: Eliminating Personal Online Information* by Michael Bazzell is a very comprehensive work by an expert in using OSINT, going far beyond privacy controls and into legal mechanisms to hide even from Marketing companies. Probably the best reference work available.

- *Open Source Intelligence Techniques: Resources for Searching and Analyzing Online Information*, also by Michael Bazzell, is the mirror image of the above work, covering the tactics and tools used to collect and analyze OSINT by investigators. Again, an excellent reference work and worth a read to understand the potential for OSINT.

---

[3] It has begun to happen. More providers now offer other options, but far from all of them.

- *The Smart Girl's Guide to Privacy: A Privacy Guide for the Rest of Us* by Violet Blue is focused on privacy for women but is useful to anyone and covers how to respond to damaging privacy breaches to mitigate fallout. Unlike Bazzell's works, this is much more focused on practical advice for everyday persons who are concerned with attacks by malicious opportunists.

In the next article, I'll be looking at threat modelling methods in a broad sense and how they are used by both designers and attackers to defend and attack systems, respectively. Specific, detailed methodologies have been defined by various groups and companies, but the high-level method and aims are fairly universal with shared goals. A quick look at attack trees, personae non grata, and the more formal STRIDE method used by Microsoft will show how they are applicable to much more than computer security.

Threat Modelling

Issue 51: 2019-12-27

For those who have been reading a while, you may know that this is the sixth article in this series which I originally planned to give a year's worth of topics. We're now at the end of that first year, and I hope you have found these articles interesting, relevant, and useful (if you have, please let the great team at Circuit Magazine know). There will be more next year, and if there are any topics in cyber you would like to see covered, again please let the Circuit team know.

We'll be looking at threat modelling this time. Threat modelling[1] is widely in use, whether knowingly or not, across every walk of life – and has been used since time immemorial to prioritise security defences. The main difference between the implicit (or explicit in some cases) predictive risk assessment carried out by everyone and threat modelling in cyber security is the attempt

---

[1] To get it over with, the US spelling is with one l, the UK spelling has two, both are valid and I tend to use the UK one, but you may see both pop up.

to document and systemise it. There are dozens of different approaches to threat modelling, and I'll briefly cover some of the most well-known[2].

> ◼ Threat modelling is a passion topic of mine, mostly because it's an incredibly useful toolset when used in combination with systems thinking, but also because I spent six months of my Master's course researching threat modelling and their grokkiness.

6.1 Seeing the Trees for the Forest

The first, and in some ways simplest, method brought into cyber security is the idea of threat or attack trees. Developed by Edward Amoroso and a collaboration between the NSA and DARPA[3], the approach is goal-oriented and works backwards. This runs a danger of being overly focused on one particular attack motivation, or having to repeat the exercise many times, but as a quick approach where you are trying to prevent attackers from reaching a single target it is challenging to beat.

Attack trees, or threat trees, are also one of the most open threat modelling techniques and the most broadly applicable, and you can see one in fig. 6.1. Some of the others we will look at (STRIDE in particular) are focused on technological threats.

With an attack tree you begin with the goal of the threat. For a basic example, we can take the goal of getting backstage at

---

[2] Shevchenko, Nataliya, *Threat Modeling*
[3] Amenaza Technologies Limited, *Attack Tree Origins*

a concert for an enthusiastic fan. Then we work backwards on what would enable them to do this, the requirements to get to that point, and so on. fig. 6.1 is a simple example of a worked tree, and the next stage would be to put mitigations in place to prevent those sub-goals (locking the staff entrance, alarming the fire escape, requiring photo ID for VIP fans, etc).

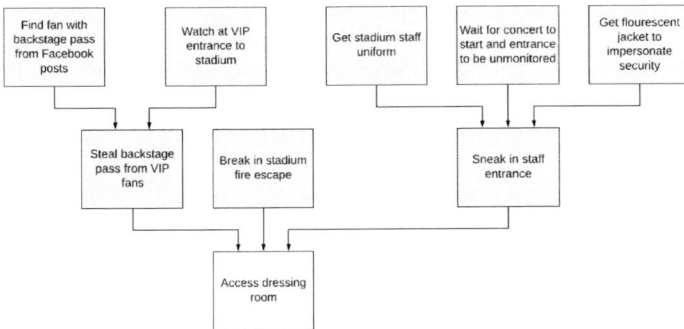

Figure 6.1: A simple attack tree example. Image credit to the author.

6.2   Personae non Grata (PnG)

A second fairly open approach looks at the exercise from the point of view of a motivated attacker. In this instance, both the goal and the means are undefined initially and instead we build a fictional hostile (or in some cases non-hostile but negligent) threat actor. These profiles can consist of anything from detailed dossiers with life histories and capability assessments, down to a few short sentences. Taking the same example as with attack trees, we would build the enthusiastic fan's profile, and then work out what goals they might try to achieve and how.

As a stand-alone exercise the PnG approach is not much use – but when combined with another model such as attack trees it narrows the options and limits the choices when brainstorming, meaning it can add significant value and help with more comprehensive, thoughtful threat modelling. With the attack tree shown earlier, the profile (fig. 6.2) might help to expand or narrow possibilities (in this case, having previously worked at the stadium a higher weighting might be given to attacks around impersonation or stealth, and some of the existing security measures could be assumed to be compromised). Of course, such profiles do not need to be fictional where a known threat can be fed into the model instead.

Andie Fan

Age: 22

Profile: Huge fan of Boys 'R' Us, and was recently fired from their job at Concert Stadium after trying to fraudulently order VIP tickets to an upcoming concert. Still has good friends who work at the stadium, and has not yet returned her access badge or work uniform. While working at Concert Stadium she had long hair, may have changed her appearance since.

Figure 6.2: An example PnG attacker profile. They can range from simpler than this, to full dossiers depending on the resources you have for threat modelling activities. Image credit to the author. Facial image generated from thispersondoesnotexist.com.

6.3   STRIDE

While STRIDE is a technical threat model designed for software development, some of the techniques involved are invaluable for any form of threat modelling. The name comes from an acronym for six attack vectors considered common in software engineering: Spoofing, Tampering, Repudiation, Information disclosure, Denial of service, Escalation of privilege. Microsoft, who developed the system, no longer strictly apply STRIDE. Other variations have been developed such as STRIPED (which reshuffles the acronym to cram Privacy in), and it is widely recognise that the categories overlap and are highly subjective. Despite this, the methods used to apply those categories are still ones I consider among the most useful for any form of formal threat modelling.

To apply STRIDE we start with a dataflow diagram – an abstraction of a software system which focuses on the way information flows through and around rather than technical detail. We then add trust boundaries, arbitrary lines designating areas of greater or lesser trust. They cannot pass through the 'blocks' of the system, and must form boundaries around them – the boundaries are shown in red on the attached diagram.

Once we have listed all our information flows (easier to do in IT than when dealing with people, but possible either way) and designated our boundaries, it's a relatively simple case of highlighting every point where an information flow crosses a trust boundary. These are our vulnerabilities, or attack points. When applying to technology we take each one and run through the STRIDE acronym, brainstorming any relevant attacks against each. Technically infeasible attacks can be listed and then discarded later – we are simply trying to be reasonably comprehensive at this point.

Once we have enough attacks against our vulnerabilities, it becomes a case of risk management against each where we need to consider the likelihood of an attacker exploiting a particular vulnerability, and the impact of them managing. While the example in Figure 6.3 is for a simple, high-level system (a smart heating system for a house) to identify the potential attack surface rather than delve into vulnerabilities, the STRIDE methodology can be a general approach that can be adapted for very complex systems as well as non-technical, physical environments.

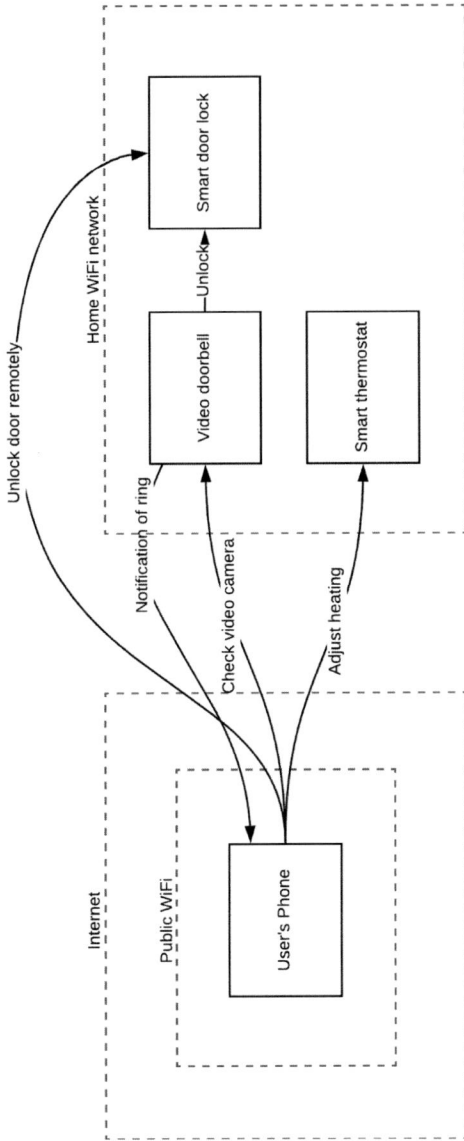

Figure 6.3: A simple example of applying the STRIDE methodology. Image credit to the author.

Threat modelling is at its most effective when it is tailored to the system being considered, and draws appropriate techniques from any source where they are useful. It also gains more value the more effort is put into it, up to a point. These are very quick, simplified examples purely to illustrate the techniques.

> ■ Since this article was written the Threat Modeling Manifesto[4] (note the US spelling) has been published. It's worth a read, but to boil it down to the fundamentals it describes four key questions:
>
> 1. What are we working on?
>
> 2. What can go wrong?
>
> 3. What are we going to do about it?
>
> 4. Did we do a good enough job?
>
> They're good questions, but since we're dealing with security which is all about people I prefer to replace the second with:
>
> 1. Who wants to stop us?
>
> 2. What can they do to stop us?

Part II

2020

Artificial Intelligence, Machine Learning, and Bias

Issue 52: 2020-02-29

Now that the introductory series is out of the way, I'll be delving more into some specialist areas of interest. The following article takes a look at artificial intelligence, particularly the machine learning area, a basic overview of how it works and the dangers of over-reliance on an algorithmic approach to analysis.

> ■ This was written before Large Language Models (LLM) became mainstream, and while they do fall under Machine Learning (ML) there's some specifics to them that aren't covered (higher dimensional spaces, vector databases, and so on). If you are interested in those specifics, then one of the unseen articles might be worth a look, despite declaring itself to be about cocktails.

7.1 What is Artificial Intelligence (AI) and ML

While the term artificial intelligence gets thrown around a lot these days, particularly by marketing specialists, there's substantial misunderstandings of what it means. Really there are two types of artificial intelligence we talk about in computing – general AI and specialised AI.

General AI is of the type that appears in science fiction rather than anything that actually exists – some experts claim it is close, but it has been 'close' now for several decades. This is generally the kind that alarmists talk about taking over the world and wiping out humans, however it's safe to say this isn't the biggest threat from AI given the remote chance of it happening outside Hollywood.

Specialist AI is the type that we have today: autonomous vehicles, facial recognition, weather prediction, and any other specialised task which a human can perform falls into this category. Within this specialist category there are broadly two different types:

Expert systems (also known as rules engines, knowledge graphs, or symbolic AI) work through coded procedures based on banks of knowledge or rules, built by taking the appropriate expert and building a flowchart of their thinking processes for a particular task.

Machine learning systems Machine learning systems modify themselves when exposed to more data, effectively learning from information they are given. This is done by setting goals for the machine, providing them with a starting algorithm or set of algorithms, and then feeding back to the machine on the success of its approach. Initially, errors will be more common than success, however as the algorithm evolves it will become more accurate. Programming a computer to play a game such as chess better than the programmer would be an example of this.

Deep learning systems[1] are not a separate category but a subset of the machine learning systems -– essentially they are multi-layered machine learning systems made up of multiple classical machine learning systems, breaking inputs into pieces and supplying those pieces to the appropriate machine learning layers before recombining to arrive at an answer.

So far so good, and the success of machine learning systems in automating work which would otherwise require significant human resources shows that there is definite value in this approach. I wouldn't be writing this article if there weren't some risks to the method though, and they come in a few forms.

7.2   The Wrong Goal

My favourite demonstration of this bias is the case of a machine learning model being trained to land an air plane (fortunately, as you will see, in simulation rather than real life)[2]. The pro-

---

[1] Large Language Models are a subset of deep learning systems.
[2] Rismani et al., *From Plane Crashes to Algorithmic Harm*

gram was given the goal of landing an aircraft with the minimum amount of force on landing. Unfortunately the simulated measurement systems had a slight fault, which meant that the system learned instead of setting the aircraft down as gently as possible, it could achieve its goal by crashing the plane with maximum force – which caused all of those measurement values to overflow and flip over to zero[3].

Any developer of a machine learning system needs to understand their goals, and how they will check them. Errors like the above are amusing only when they happen in a simulation – in real life it would be an entirely different matter.

> ■ This is one reason why diversity is vital to teams building these systems – it is incredibly easy to overlook or be unaware of something that has never affected you. When being codified into AI systems the potential for harm is huge and so biases and assumptions **must** be noticed and questioned.

7.3 Bias

Another, more insidious example, is the bias found in facial recognition systems. Because the data fed to these machines tends to be weighted heavily towards a particular set of features (largely North American Caucasians), they have trouble identifying any other set of features reliably. Anyone who does not fit into the category the machine is expecting is at much higher

---

[3] Buffer overflows like this are a fairly common method of exploiting systems.

risk of being misidentified – given that facial recognition is used widely for security and law enforcement purposes the problem here should be clear.

> ■ This failure to identify feature sets outside those used for training you would expect to be easily resolved.
> It still happens. If anything it's worsened with generative AI presenting even more visible biases when creating images.

Worse still is the case of algorithms used to assess a defendant's likelihood of recidivism. ProPublica[4] investigated one particular tool, comparing the predicted recidivism rate for white and black defendants against the actual rate. Broadly speaking the tool was correct about 6 times in 10, but the devil is in the details. White defendants were more likely to have their risk of recidivism underestimated. Black defendants were more likely to have their risk of recidivism overestimated. This was not a minor error, as white re-offenders were mistakenly labelled at low risk twice as often as black re-offenders, while black defendants were incorrectly labelled as high risk twice as often as their white counterparts. The picture becomes dramatically worse when you consider that in terms of violent recidivism (given a separate rating) the algorithm was only correct with its predictions 20 percent of the time.

One of the problems when these biases emerge is that it is extremely difficult to understand without relying on the system

---

[4] ProPublica, *Machine Bias*

and checking the validity of predictions. In the case of the recidivism tool, this was the product of two years of research covering over 100,000 cases. With the speed that we are developing and applying machine learning to solve problems, a lot of damage can be done before any systemic biases are uncovered — if they ever are.

> ■ Even when systemic biases are revealed and well documented, it may not make any difference. Despite repeated demonstrations of its biases, the COMPAS algorithm mentioned above is still in use to advise on parole decisions in some US states.

7.4   Black Box Systems

Complicating this further is that once a ML model is built, its decisions cannot usually be explained by the developer. For complicated problems the process of learning results in what are, to all intents and purposes, black box systems[5]. The algorithms grow so complex that they are simply not understood by those who set the initial conditions, meaning if an unexpected input occurs outside the model the output is unpredictable, and we cannot analyse why a decision was made.

The essential lesson is that machine learning is just providing statistical information, and just like any information it can be wrong. While they are capable of sorting more data more quickly

---

[5] ScaDS_PubRel, *ScaDS.AI - Center for Scalable Data Analytics and Artificial Intelligence*

than a human, and are arguably less prone to erroneous decisions given the right data, they do make mistakes and taking them as a source of ultimate truth magnifies the impact of those mistakes. Be sceptical of any decisions from machine learning systems as you would from any human expert, and whenever possible validate the output.

Introduction to Steganography

Issue 53: 2020-04-30

Cryptography, which includes steganography, is a particular interest of mine. While nowadays most of the introductory codes are useful for understanding fundamentals, the mathematics involved for modern cryptography lend it more towards advanced courses and specialists. I didn't think it was worthwhile going into the Caesar Cipher, or mechanics of the Enigma code, in this article. Steganography is still very relevant today and is arguably the hardest form of cryptography to computationally detect and crack.

> ■ Cryptography is worth going into more, simply because it's fascinating. There are excellent **books** available, including a suggestion in Further Reading.

8.1 What is steganography?

Cryptography derives from the Greek words *kryptos* (hidden, or secret) and *graphein* (writing). It includes a whole host of techniques, and one which stands out is steganography which comes from the Greek *steganos*, or covered[1]. What makes steganography unique is that it is a technique of making information hard to find, not hard to read once found. There's a common phrase in cyber security – 'security by obscurity is no security at all', and the concept goes back to 1851. Steganography challenges that idea by providing security only through obscurity.

Steganography is not so much a single technique, as a family of them, with new ones being added all the time. All of the techniques share one core concept, trying to hide a message. I've heard compelling arguments that Renaissance artists using symbolism in their works were practising a form of steganography, and ideas like the language of flowers go back centuries. When children draw stick figures using semaphore as secret messages to each other, they are practising steganography.

Modern steganography is dramatically more practical, and usually a lot more malicious, and comes in a few fundamental forms. It is used in malware command and control, data exfiltration, and exchange of illicit information and material. If you do not know where to look, or what to look at it is frighteningly hard to detect. When the hidden message is effectively encrypted, there is very little that can be done.

Steganographic messages can be hidden in audio, video, or im-

---

[1] As does *stegasaurus*.

ages with simple, free tools that can be downloaded and run on a mobile phone by anyone. While it is an inefficient method, as the carrier message must be significantly larger than the hidden message, in today's high-bandwidth world of social media that inefficiency is not an issue. A steganographic message for exfiltration of data can also be hidden through tunnelling different protocols – a popular method being the use of the domain name system (DNS) queries, allowed through almost all firewalls, to exfiltrate data or infiltrate command and control messages into an existing infection. As the internet as we know it relies almost completely on DNS to work, blocking this is challenging. Methods do exist to detect and protect against this vector, but they are hardly ever deployed.

First we'll look at the more human side, with embedding messages into media files.

8.2   Steganography Tools

One of the best ways to understand the human side of steganography is to try out some steganography tools. Steghide is one open source tool which will encode a message into almost any media file you care to name, protected by a passphrase and an additional layer of encryption. At the recipient end, or on your own system, you can extract the data equally easily. Pixelknot for Android, and Pictography for iPhone, are similar smartphone-based systems.

While steganography does have a sinister side, the difficulty of detecting it does make it highly suitable for a lot of benevolent purposes. Any time someone is under electronic surveillance, if they have a pattern of activity including social media postings and a previously established protocol, uploading a selfie with a steganographically encoded message is a simple, fairly secure, and quick way to communicate. This has been used in practice across the world, including for journalistic reporting from areas with surveillance-heavy authorities. In fact Pixelknot, the Android steganography tool, was developed as part of the Guardian project as part of their mission to support activists, journalists, and humanitarian organisations.

8.3   Malware and Steganography

One of the biggest problems for sophisticated malware is communications – it is when reaching out to, or receiving messages from, Command and Control (C2) systems that malware infections are easiest to detect and at their most vulnerable to disruption. Many of the largest botnets taken down have been disrupted in exactly this way, with white hats detecting the method for contacting their C2 infrastructure and either compromising it in turn (then sending out a self-destruct message, which is not as cinematic as it sounds), or breaking it through other means[2].

The most sophisticated modern malware makes heavy use of steganography not just for C2 purposes, it is also used to exfiltrate bulk data. The C2 side can vary from social media post-

---

[2] Jedrzej Bieniasz and Krzysztof Szczypiorski, *Steganography Techniques for Command and Control (C2) Channels*

ings[3] through tunnelling protocols. It's the tunnelling protocols we'll take a brief look at now.

8.4 Tunnelling Protocols

At a basic level a tunnelling protocol allows data to be sent from one network to another. That data can in turn be a tunnelling protocol. There are some amusing implementations of this, for example the underlying protocol that provides most modern networking, TCP/IP, has been implemented using Facebook chat (which gave it very high latency and low reliability), and carrier pigeon (high latency, medium reliability, huge bandwidth).

Domain Name System (DNS) tunnelling is the most common and well known steganographic tunnelling method used by threat actors. DNS works through queries sent up a hierarchy of servers to resolve domain names to server or service addresses, and these queries are forwarded as needed. To use DNS tunnelling an attacker does not need any special relays inside a network, these are all provided as part of the legitimate network infrastructure. All that is needed is a malicious authoritative server for a domain or domains − queries are then sent for that domain, carrying the data as part of the query. The server receiving the query will then reply with a return message − and any networking protocol can be encoded through this. Tools are available for DNS tunnelling not only for data exfiltration, but instant messaging, video conferencing, and almost any other protocol that is available normally.

---

[3] Twitter is fairly popular for this, and I suspect there will be campaigns picked up on Instagram before long.

In 2016 Infoblox[4] found 40% of malicious software they tested made some use of DNS tunnelling – in the years since this will likely have grown as open source and off-the-shelf toolkits have become available. Highly sophisticated attacks, including ones suspected to be sponsored by nation states, use DNS tunnelling for data exfiltration and C2. It is perfectly possible to detect and/or prevent DNS tunnelling, but it is part of the suite of cyber security hygiene measures that are very rarely implemented due to lack of resource, lack of funds, lack of awareness, or a combination of the three.

---

[4] Infoblox, *Nearly Half of Enterprise Networks Show Evidence of DNS Tunneling*

Social Media Engineering

Issue 54: 2020-07-13

A previous article, the second in last year's series, spoke about social engineering as a way to manipulate individuals into disclosing information which would benefit an attacker. I'm sure recent events have been obvious to everyone, and that most readers will have been aware of some of the conspiracy theories, dis- and misinformation, and polarisation which is occurring. While this is, in part, down to the way that certain social media platforms[1] are designed to focus on division, some of it is down to deliberate recruitment techniques. These techniques are universal, and are often used to recruit people into personality cults, Multi-Level Marketing (MLM) schemes, conspiracy theories, and various other radical groups. They are low-cost, low-effort, incredibly effective, and even more effective in an environment where most social interaction has been moved on-

---

[1] Pretty much all of them are designed this way. Emotion creates engagement, the easiest and most effective emotion to trigger through social media content is anger, and so almost all of the algorithms promote divisive content.

line and face to face is discouraged. They are used to help spread conspiracy theories which target and encourage action against critical national infrastructure, and high-profile persons.

To take just one example the conspiracy theories around 5G were quickly re-tailored to incorporate Bill Gates and George Soros as targets, and were effective enough that there were arson attacks against 5G infrastructure. In other words, a few memes and videos, along with the use of these techniques (whether deliberate, or accidental, but there is strong evidence of deliberate action) were enough to cause home-grown domestic terrorism attacks on UK soil[2].

■ Heading into 2024 I have enjoyed the conspiracy theory associated with the push towards Electric Vehicle (EV)s being down to sinister authorities wanting to disable cars at the touch of a button. To reassure anyone reading, the technology to disable cars at the touch of a button has been around for decades, and it is no easier or harder to remotely disable an EV than an Internal Combustion Engine (ICE). You can read more about it in the article on self-driving cars.

While not part of what might normally be considered cyber security, social engineering is usually included, and a lot of these techniques exploit botnets and catfish accounts (fake social media accounts designed to present an appealing persona in order to infiltrate social networks) to encourage their spread. This

---

[2] Meese, Frith, and Wilken, "COVID-19, 5G Conspiracies and Infrastructural Futures"

technique, and others like it, show the weaponization of social media.

9.1   Recruitment Steps

Classically, recruitment techniques used by cults or similar rely on four basic steps. We'll do a quick run-through of these before looking at how they can be modified to thrive on social media platforms[3].

Target selection

> The right target is often someone who has suffered a crisis. Any identity-threatening crisis such as grief, significant life change, or a global pandemic will do. All of these prime someone as a potential target as people often reassess identity following significant change and trauma. Graduates, and recent arrivals at universities, are often favourites as these are very public, very obvious events which will prompt identity reassessments, along with the financial situation of many students leaving them vulnerable.

Love bomb [4]

> Simple and straightforward positive reinforcement. The recruiting group showers the target with compliments and support, simply being nice. This support is unconditional,

---

[3] Matt Davis, *4 Psychological Techniques Cults Use to Recruit Members - Big Think*

[4] Dale Archer, *The Danger of Manipulative Love-Bombing in a Relationship | Psychology Today*

with no caveats or constructive criticisms, and when it works it positions the group as a supportive best friend.

Isolation

After the supportive position has been established, the target is isolated. This is one of the areas that adapts well to social media, and we'll see why shortly. At this point with the target free from outside influences, a whole slew of emotional manipulative techniques can be applied to transition the group from a supportive best friend to a core part of the target's identity.

Maintaining control

This step most replicates abusive relationships, with the group being a source simultaneously of support and terror. The threat of withdrawing the support allows much more obvious use of abusive tactics, and by this point if the other steps have been successful the target will have started to think of themselves as a member of the group, as a core part of their identity. Breaking away, as with any abusive relationship, is difficult, painful, and usually requires outside help and we will examine how this plays into social media engineering as well.

9.2   Translating to Social Media

In person these techniques are clearly recognisable to anyone familiar with this sort of recruitment. Even some less-ethical sales books promote similar methods (MLM companies are founded on this sort of psychological manipulation). Where it can be less obvious is when they've moved to social media, where the steps are somewhat different.

9.3   Target Selection

With the breadth of social media, targets often self-select. A few years ago it was established that 64% of members joining extremist groups on Facebook, of all kinds, came from recommendations from the Facebook algorithm. You may have come across the term 'dog-whistle politics', meaning phrases or ideas that seem innocuous, but carry greater meaning for a particular audience (if you read the article on steganography, dog-whistle politics is a form of steganography making use of what's known as a context code).

By interacting with the post, whether responding to the hidden message or not, someone is marking out that they are a potential target. Because these techniques require a much lower investment of time and effort on social media than in person, a much wider net can be cast and so targets can be selected with less (or no) care beyond clicking 'Like'.

9.4   Love Bombing, Isolation and Maintaining Control

The social media approach combines these two into a single step. Whether it's through a dog-whistle post of some kind (often a meme), through sharing news from extremist and unreliable sources or groups, or through some other mechanism, the aim is to get the target to post something controversial themselves. At this point people who are aware of the dog-whistle context and disagree with it will often react in a knee-jerk fashion with shock and anger. Not only because they disagree with the view, but because of the psychological principle that everyone considers themselves at the centre of a story – so a post that's ideologically against a person feels like a personal betrayal.

Of course that anger leads to a reaction from the poster, who feels attacked because they meant no harm. Compounding this, the group who originally created the innocuous post will jump onto the conversation, giving the unconditional support and acclaim, while friends and family are more likely to react with negative emotions. This is devastatingly effective – trimmed back to the most shallow level social media is built for instant, knee-jerk responses rather than nuanced discussion. Emotionally weighted reactions can very quickly alienate a target from their friends and family, while simultaneously giving them a supportive new network among the radical group.

> ■ This has become increasingly common as social media causes greater and greater polarisation of views, and the tidal wave of content shortens everyone's attention span to digest and deal with nuances. Whether done intentionally, or unconsciously through normal group cohesion dynamics it can be just as damaging.

This reoccurs as the target will tend to then share more controversial posts – both as they get drawn into the group and to upset former friends and family in revenge for their perceived attacks. It merges perfectly from this stage into the keeping control stage, as a person is rejected by wider networks for extreme views they are driven more towards those perspectives.

9.5   Countermeasures

The best countermeasure is awareness and education on how these techniques work. Another trait is that people do not enjoy the idea they are being manipulated, and effective education works. Once a target has begun re-identifying as part of one of these groups it is a much more challenging and complex problem, and while de-radicalisation programmes exist they are very much still in their infancy.

> ■ Whether or not de-radicalisation programmes have improved is debatable. The UK's approach is largely centred on Prevent[5]which pushes a lot of the responsibility onto unqualified public sector workers such as education and healthcare professionals, who go through about an hour of online training to learn how to recognise and prevent people from becoming radicalised. Draw your own conclusions on the effectiveness of this approach.

---

[5] *Prevent Duty Guidance*

Weaponising Social Media

Issue 55: 2020-08-31

Last issue we spoke about techniques used on social media to isolate and recruit individuals. This time we're going to be looking more at how conspiracy theories and opportunistic misinformation is misused and amplified to break down trust in authorities and experts in order to cause harm. Often these misinformation attacks end up focusing on particular individuals as being behind shadowy conspiracies.

The psychology behind the misinformation was mostly covered last time, so we're going to look more at the mechanics that enable them. We'll take a look at fake social media accounts, botnets, and automated amplification to exploit social media algorithms.

10.1   Botnets

Botnets consist of computers, or other devices (the Mirai botnet, one of the largest known for a time, ran on CCTV cameras), compromised and enrolled into a central control system. While one of the most common usages is generating traffic for Distributed Denial of Service (DDoS) attacks, combined with a bit of scripting they can be used to generate and control fake social media profiles with a much lower risk of detection. Using multiple devices in a botnet means that true source IP addresses are masked, and one of the easiest ways to detect coordinated creation of fake profiles is checking if they have the same source.

In the simplest form this sort of generation scripts the manual steps that would be taken to generate a profile, creating a virtual browser, clicking buttons, and entering information. Fortunately, less effort is spent on these than could be, leading to some common weaknesses.

10.2   Fake Profiles

It takes a few minutes' work to manually set up a fake social media profile, and while there have been very limited improvements to filtering systems none of them really do much to prevent it. Verification via phone number or e-mail isn't effective, as it's trivial to create a burner number or e-mail address for an account. One of the most effective checks used to be a reverse image search on profile pictures, but with the emergence of services like thispersondoesnotexist.com that's no longer an option.

> ■ Twitter, now stuck with the label 'X', claims its premium subscription prevents fake profiles. There is little evidence that this is the case, especially given the number of bots which have premium subscriptions. Simply put, the benefit from these is enough to justify the cost for the people using them.

Of course what can be carried out manually is much faster to automate. With some effort put into scripting it's possible to generate hundreds of fake profiles in minutes, using different source IP addresses, auto-generated e-mail addresses for verification, and some username generation. This is where one of the weaknesses comes in and you will occasionally hear, particularly on Twitter[1], comments that usernames followed by strings of numbers tend to be bots. The mistake that can be made is assuming this means they are entirely run through scripts and automation – they are generated by botnets and scripting, but often at least partly run manually.

10.3 Amplification

Social media application interfaces are used to monitor and reply to comments, as well as broadcasting messages. Because the troll farms (a term referring to buildings where these campaigns are coordinated) control so many accounts they also use

---

[1] I considered updating this to X, but it's my book and I can be nostalgic and petty if I want to.

them to rebroadcast posts which match whatever narrative they are trying to develop. This activity is combined with combing various sources for new disinformation to throw into the mix, sowing more confusion and furthering the agenda of distrust[2].

As you'd expect, one of the targets for this disinformation is denying the existence of troll farms, and discounting claims of their existence as conspiracy theories. The effect of this is easy to see – when there is a lot of disinformation around, throwing more into the mix just increases mistrust. If a coordinated campaign can convince people to be equally sceptical of all information, finding the accurate information becomes as much a matter of chance as anything else. The trick then becomes simply filling people's feeds with as much disinformation as possible – if you can crowd out reliable information and overwhelm people's ability to critically think through sources, the system breaks down.

10.4   Targeting Individuals

Where this gets dangerous to a high-profile individual is where those who are convinced by this information are easy to recruit. With the human talent for pattern recognition, these scattered opportunistic pieces of disinformation are tied together into meta-conspiracies which are then pinned on individuals. The people who are convinced by these meta-conspiracies are demonstrably dangerous – last time I mentioned arson attacks on 5G towers, but there plentiful documented incidents of these

---

[2] *UK Exposes Sick Russian Troll Factory Plaguing Social Media with Kremlin Propaganda*

theories inspiring domestic terrorism, including bombings and shootings, in various countries[3].

> ▰ In a semi-related case a man who attempted to kill Queen Elizabeth II was encouraged to do so by his 'girlfriend'. The girlfriend was an AI chatbot. You can read all about this in the article on parasocial relationships.

Recent examples include these theories targeting heads of state (a reinvention of David Icke's shape-shifting lizards controlling the world conspiracy), Bill Gates and George Soros (microchips in vaccines), and many others. Alongside the use of a fake social media event to co-ordinate an armed militia of 200 people in America recently, the weaponization of social media is a threat which anyone providing protection services for a high-profile individual needs to seriously consider.

10.5   Protection

Prevention of this sort of weaponization relies on education and effective action by social media platforms. Education is not effectively scalable and cannot be comprehensively deployed, and social media companies have not shown any indication that they

---

[3] Pema Levy, *Facebook Groups Sent Armed Vigilantes to Kenosha. Your Polling Place Could Be Next. – Mother Jones*

can address the threat effectively[4]. Recent attempts such as Facebook's striking off of a large number of accounts run by a Romanian troll farm are a good sign, but barely make a dent in the problem.

A good threat intelligence programme is needed to protect against this kind of threat, making sure that any such attempt to weaponise social media against an individual is detected as early as possible so precautions can be taken. Some effort can be made to counteract the claims, but denial is rarely effective against these movements. The only advantage of the rapid development and deployment of these is that the attention span is often short term, so the threat can develop from nothing unexpectedly, but if precautions are taken is likely to also subside quickly.

10.6   Further Reading

If you are interested in these threats and would like to read more, then there are a few different sources. The Internet Research Agency is the best documented case, *Active Measures* by Thomas Rid is one of the better books on the subject, and it's an area where more and more threat intelligence providers are sharing research on the subject.

---

[4] More importantly than their lack of competence in addressing the threat, documents have revealed that at least in some cases they have been well aware of it for some time and have chosen not to act since it would threaten profits.

Emotional Phishing

Issue 56: 2020-11-30

We've spoken a couple of times now about tactics used to weaponise social media, and you will have spotted a pattern in these. As with many methods used to manipulate they do not try to argue someone around to a position in a rational manner, instead they are designed to cause and exploit emotional responses.

Phishing is a whole family of different attacks designed to manipulate a target into performing actions against their best interest, usually disclosing information in some way or carrying out fraudulent transactions. Within that family we have vish (voice phishing, over a phone call), SMish (phishing via SMS, or text message), business email compromise (also known as BEC, phishing from the pretext of being a senior executive in a business), spear phishings (highly targeted and personal phishing), whaling (highly targeted phishing against an executive), and many others.

The overwhelming majority of cyber attacks occur through some form of phish. The recent, well-publicised, compromise of numerous verified high-profile twitter accounts was enabled through a form of phishing rather than a technical exploit.

As an industry I am sorry to say that cyber security struggles to engage with people to prevent phishing. Doubtless you'll be familiar with the annual phishing training approach, with a quiz at the end, telling you to look for typos, suspicious links, and similar. There are also, of course, 'practice' phishing campaigns you can engage which are usually a little more effective, but still don't protect everyone. Generally these campaigns, over 12 months of continuous awareness and training, can reduce response rates to a few percent of employees. The problem is that this lower response rate doesn't last unless the training is continued, and can start to climb again as people become complacent.

What almost every attempt at training I have seen gets wrong is that they try to highlight small technical inconsistencies in phishing e-mails as the flaws. Not only will a well-crafted phishing attempt display none of these, but they are very easy to overlook in the moment. I was recently asked to give a keynote on this issue by a company, and it's some of the information delivered during that which I'm going to share now to help you understand why the technical phishing training doesn't work as well as expected, and how you can give people the behaviours and tools they need to protect themselves.

> ▪ Over the last couple of years my company has started to offer this form of training as a service. It is effective, but it does need to be maintained – and importantly needs to be refreshed constantly. I stand by the numbers above: you can reduce the response rate to a few percent at best, and you have to maintain it. Even then, teaching people to develop the pause habit in response to any emotional punch is equally effective, while both in combination are even better.

11.1   Phishing is Personal

Even when phishing is a volume attack, targeted against everyone, it feels extremely personal. With the number of data breaches it is almost guaranteed that everyone has at least one old password out there (check Have I Been Pwned with your e-mail addresses to see what I mean). Bad actors take advantage of this to inject an extra personal angle into their attacks, even with modern mass attacks as the picture shows.

Even where the attack is intended to take money from a company rather than an individual, it will still be intended to exploit the human recipient, not an abstract corporate entity. The people behind these attacks are only too well aware that ultimately it comes down to the person to achieve their goals.

11.2   Phishing is Emotional

A phishing attack is not intended to convince the recipient through logical, sensible argument to do what the attacker wants. It is usually not intended to fool someone with perfect construction of the message. Instead a phishing attack is meant to bypass all of that careful training, the list of rational checks to make for any e-mail, and get straight to the heart of the matter (metaphorically speaking). A phishing attack is almost always an attempt to exploit emotions, whether that's to trigger greed over an offer, a sense of fearful urgency because an authority wants something done, hollow dread at threats of public embarrassment, or panic due to some bureaucratic hiccup.

There are two basic behaviours which can protect against the vast majority of attacks, without digging through text looking for typos.

The first is to both trust and question your gut. If an e-mail, or any other communication, triggers an emotional response then the automatic behaviour should be to take a step away from the communication and breathe. Once calm, start questioning the message, and the best way to do so is to contact the sender through some different, trusted means. If it's supposedly a message from a bank, phone the bank with the number on your card rather than the one in the e-mail. If it's an e-mail from a senior executive, get in touch with them directly, or their assistant, with details that are not in the e-mail. Better of course is to contact your security team with details, but there are instances where a company won't have a team to contact.

The second is to question anything the first time. If anyone is asking for something new, a change from the normal way, then as with the emotional response above use some other trusted channel to contact them.

These two behaviours will not protect against certain attack vectors (invoice fraud can be devastatingly effective when well-crafted), but they are two of the most effective behaviours to train to protect yourself and anyone else.

Part III

2021

Cyber Security and Humans

Issue 57: 2021-02-01

We're now into a new year, and the third year's run of these articles.

Over 2020 cyber security and technology have only soared in terms of profile and importance, with talk about threats to remote working from technology, difficulties, and some dramatic outages. Logistics, enabled largely by technology, have been essential to keep things moving and give people support and normality.

We've also heard a lot about attacks on medical research centres, looking to access research information and believed to be carried out by nation states. We've seen attacks, traced back over multiple months, which have penetrated organisations to the deepest levels via their suppliers with the SolarWinds compromises.

Over the year, I'll be looking in depth at some of these incidents as more details come out. In a timely fashion, I wanted to start the year off with the biggest of these, which is still ongoing, why it is far from the last, and what can be done about them.

12.1 SolarWinds Orion

SolarWinds make a suite of products aimed at helping organisations manage their technology. One of these is called SolarWinds Orion, which provides a dashboard and management interface for different technology environments. Referred to as a 'single pane of glass', it allows you to view and manage your physical on-premise technology, cloud environments, and the mixed environments, all from one place.

It's simply not possible to manage everything without either a ten to a hundredfold increase in people for an IT team, or a management solution like SolarWinds. To manage these servers, that solution needs to have highly privileged access to the servers. It also needs to be able to deploy software, update systems, and change their configuration, all from one central point.

Building software to do this safely and securely is a significant task, so it is extremely rare for companies to build their own. The Orion solution by SolarWinds was used by over 30,000 private and public sector customers to manage their networks. It was not always the only solution in place, but even where it was one of many it would be used to control a significant part of the network.

12.2   What Happened?

In early December cyber security company FireEye was breached through, at the time, unknown means. Their library of attack tools for penetration testing, mostly well-known exploits, was stolen. While embarrassing and, for FireEye, inconvenient as their arsenal of custom pen testing tools was now essentially useless after they had to release details, initially this did not seem like anything more than a single target.

Three days later FireEye announced that they had uncovered a much larger breach, with a component in the SolarWinds Orion software having been altered for malicious purposes sometime between March and June, and rolled out through SolarWinds own automatic update servers.

The malicious component provided a backdoor for an attacker group codenamed UNC2452 by FireEye to control the software. Since Orion controlled servers, the result was that the attacking group potentially had unfettered access to any network making use of the Orion software[1].

At this point we can make some guesses about the motives of the attackers. If it was an organised crime group, then as with the breach on Twitter it would likely have been a fairly short-lived attack in which ransomware or similar was deployed to cause as much chaos as possible and raise funds. Since in- stead the attackers have since been found to have moved slowly and carefully, identifying valuable targets – particularly intelli- gence targets – and maintained the compromise until FireEye

---

[1]   Temple-Raston, "A 'Worst Nightmare' Cyberattack"

discovered them after the leak of their tools, it is a reasonable assumption that an intelligence agency not motivated by a profit agenda is at fault. FireEye's UNC2452 group is known by the US government as APT29, and among other names is codenamed Cozy Bear[2]. Any threat codenamed Bear is believed to be associated with Russia, and Cozy Bear have been involved with a number of other attacks including accusations in July of attempting to steal data on vaccines at treatments, made by the US NSA, UK's NCSC, and Canada's CSE.

Importantly, while Cozy Bear are a well-resourced, sophisticated threat, the main difference between a nation-state level attacker and an organised crime group is not the level of capability, but the motivation. The compromise of SolarWinds was sophisticated, but none of the attack vectors used were new in principle – they are well understood attacks which can be addressed by an effective security programme, and mitigated or prevented by effective security by design.

> ■ To be clearer, the compromise itself was due to a bad password. The exploitation of that compromise was highly sophisticated in injecting malicious code undetected into a product update.

---

[2] Yes, we really need to improve naming conventions for Advanced Persistent Threat (APT) groups.

12.3   Supply Chain

While there are understandable concerns about Cozy Bear, there are serious concerns around supply chain security as well. The supply chain has been considered a valuable attack vector by some in cyber security for a while, and one worthy of attention, but the difficulty of effectively assessing the security of the chain and the risk for any individual company means it is often either overlooked or given only cursory due diligence.

At that point organisations have no choice but to trust their suppliers. When those suppliers are trusted to have all of the access and privileges of the most senior, highly-permissioned administrators it is stunning that most organisations put more effort into background checks on selecting their own trusted staff than into the supply chain.

The supply chain, obviously, is not a new threat and is often discussed in physical security arenas. The difference is that for any non-software supply chain there is a limit to the impact of a breach. Devices with malicious hardware installed can at least only affect new installations, while with automatic updates the injection of malicious software can affect not only new installations, but all of those which already exist.

The SolarWinds breach is going to be a long saga – at the moment I'm aware of a list of 250 organisations (including government agencies) confirmed as affected, and there is a much longer list of those potentially affected. In my next article I plan to write about the importance of incident response exercises – usually carried out by organisations no more than once a year at most – and how they can be run by anyone to help pull out the holes in a security framework or response plan, as well as used for training incident response teams and associated staff in how to deal with these attacks.

> ■ The guess it would be a long saga was accurate. While there's less rumbling about it now the reason given by the company was that it was down to a bad password set by an intern. Blaming the intern has lived on as a running joke in cyber security ever since.

cyber **Security**

Issue 58: 2021-03-31

> ■ The above was not and is not some sort of printing error or typo, it's a clever representation of the topic of this article.

I write and speak about this particular issue quite regularly, and it's one that I believe is vital to grasping cyber security's place in the world. Especially while many people are still remote, technology has become more and more central to people's lives, and we are talking about ways that things will or will not return to normal. As we hear about more and more cyber security incidents, each supposedly carried out by 'sophisticated threat actors with unprecedented capabilities', it's time to talk about the mystique of cyber security and the problem it has with public perception.

Jon Moss once said to me, when I asked him for a definition of security, that it is the art (or science, it's been a while and I forget which) of protecting an asset from a threat. In many security fields that is immediately clear and obvious to practitioners. In cyber security, information security, or IT security it can be muddied and hidden away. Since it's been a while since I last wrote about this topic in the Circuit, it's time to dust off the cobwebs and reiterate some things, as not much has changed in the field since the last time I brought it up.

13.1 Cyber security is not magic

There is an incredibly common perception, encouraged by some cyber security professionals and companies, that cyber security requires some sort of arcane, obscure, special knowledge which only a privileged few can access. This perception not only discourages people from entering the field and taking ownership of their own security, it also gives an impression that cyber is somehow outside the reach of anyone other than specialists.

> ■ Every time I think things might be improving, someone proves me wrong. Recently I've seen people comparing entering cyber security to becoming a brain surgeon. These are people who have influence in the industry, and they put out this kind of nonsense. I can't see this problem going away any time soon, but there are at least more people arguing against it.

With the media stories out there, thinking about cyber security is stressful for many and the promotion of this view drives learned helplessness. Learned helplessness[1] is what hap-

---

[1] Maier and Seligman, "Learned Helplessness"

pens when people repeatedly experience a stressful situation and feel it is out of their ability to control. After enough experiences like this, which doesn't take many repeats, people stop trying to do anything even when an opportunity to change arises. Getting people out of learned helplessness is difficult, and for years much of the cyber security industry and the media coverage has been driving the idea that not only is everyone under threat, but that protecting yourself from those threats is not possible without abilities beyond the reach of ordinary people.

This idea extends to security professionals as well. Over the years I've had several conversations with experts in various security fields who are convinced that while they have expert knowledge, far beyond mine, in the security discipline they need to leave cyber security to the specialists.

13.2   It's a domain, not a discipline

Security is a discipline, a skills toolkit, more about learning how to approach situations in general than about the details of those situations. The skills involved in security are applicable across multiple different domains, and all in the pursuit of protecting assets from threats. Cyber is a particular domain, an area in which those skills can be applied. All of the skills developed in other domains of security can be applied to cyber security by learning to reapply the models you use to technology.

All of this is to say that cyber security is just security applied to a poorly-defined mishmash of technology and information security. It is not special. It does not deserve to be placed in its own ivory tower. Anyone with the right level of curiousity can not only learn it but excel if given the opportunity. There's no need to have amazing technical skills, just an understanding of how the technology can be used and what attack vectors might exist.

Sure, the technical skills are useful, but they aren't essential for individuals to take ownership of their own security, or to protect other people. Call in the specialists when they're needed, but take the time outside of that to ask questions and learn.

Any cyber security professional who won't help people to understand the field most likely doesn't understand it themselves. To improve cyber security worldwide it isn't enough to add new people to the field, we aren't ever going to have enough and we struggle to get companies to understand what security they actually need in any case (if you're wondering for the vast majority of companies the answers are 'more than you have', along with 'fewer shiny technologies and more people'). We need for everyone to feel comfortable when dealing with a cyber security situation, to not suffer from learned helplessness but instead to take control of their own security posture, take responsibility for their own protective measures, and ask for help where it's needed.

Ransomware, Insurance, and Backups

Issue 59: 2021-06-17

Most people are not aware of the real impact and threat of ransomware campaigns going on almost constantly. While I'm sure everyone knows of the ransomware attack on the Colonial pipeline, and the more recent similar attack on JBS, these are only the latest and most visible in a continuous stream of attacks. These are the tip of an iceberg, and the true scale of the iceberg is hidden not only by the media stories, but also by the reluctance to really talk about the problem.

In the Colonial pipeline case[1], a ransom payment was made of $4.4 million. It sounds sizeable until you look at the bigger picture. Estimates of the worth of ransomware as a global industry range between $1 to $10 trillion. That means if every single ransomware attack received the same payment as the Colonial pipeline, we would be looking at a minimum of 200,000 such attacks each year, more than 500 per day. The vast majority of

---

[1] Sean Michael Kerner, *Colonial Pipeline Hack Explained: Everything You Need to Know*

attacks do not even come close to that level of cost, so we are looking at a lot more incidents, impacting a huge number of people every single day.

> ▬ The estimate here is still repeated, but finding a reliable source for it is challenging. It appears some reports, and some marketing companies, have either deliberately or accidentally used ransomware and cyber crime interchangeably. Ransomware is a subset of cyber crime, which includes everything from fraud to intellectual property theft, and the multi-trillion estimate is usually associated with cyber crime as a whole.

14.1   So what is ransomware?

While the general view of ransomware is software that encrypts some files then demands payments to decrypt them, modern ransomware organisations have developed their methods somewhat. The encryption of files still happens, however alongside that attackers will often exfiltrate information and threaten to publish it if payments are not made. Partial publications happen to back up the threat, and honestly there is no guarantee that on payment of the ransom the victim will receive their own data back, let alone prevent future threats and publication. What is guaranteed is that they are now marked as a potential income source for future attacks.

It's also important to note that much of today's ransomware is effectively a commercial product. Criminals can license the software, or even purchase it as a managed service, and deploy it where they've already gained access. Alternatively and growing more and more common a group may license the core software, customise it, and buy access to organisations from access brokers who make their own living simply by finding ways in and selling them.

Even where a payment is made and the keys are provided to decrypt the files, there are no guarantees that they will work any faster than restoring from backups, if they work at all.

14.2   What are the good guys doing?

Many security researchers look for ways to break or exploit ransomware, finding ways to decrypt files that don't require any communication with the attacker. This is an ongoing battle, and the debate still rages about the publication of these tools or keeping them quiet. In a recent case a certain well-known security company published their decryption tool, using a hole in the software discovered by a researcher. Of course the ransomware organisations have access to the same internet as the rest of us, found the tool, reverse engineered it, and improved their software to prevent the hole from working in future.

14.3   What's the real impact?

Cases like the Colonial and JBS attacks are well-publicised, and relatively rare. The vast majority of attacks never reach the headlines, and just as with everything else in business most of them don't affect such high-profile organisations.

In 2017 it's estimated that one third of small businesses world-wide were affected by ransomware. Of these one fifth had to cease operations. Not a temporary stoppage as Colonial and JBS with insurance and reserves to get through the incident enacted, but permanently closing their doors. Most of these attacks succeed because of human error, not a clever technical vulnerability. From what we know of the Colonial attack, like the SolarWinds attack which has largely been forgotten but ended up giving the attackers root access to government agencies and military research organisations across the world, it was down to a poor password choice.

> ■ I've left these numbers in, because there are sources to support them, but in the time since I've done digging into how these numbers are sourced. The simple answer is that there are no reliable, trustable numbers on these topics. Anyone who quotes statistics about how bad things are as a way to sell their product has almost certainly come up with their own numbers as the reliable research simply does not exist.

Many businesses which haven't been impacted by ransomware believe that they aren't a target, and even some who have been subject to small incidents assume that they will not be affected by anything larger, despite often not even understanding how the malware got onto their systems in the first place.

14.4   What can businesses or individuals do?

There are two parts to dealing with ransomware attacks. The first is prevention, and often that's simply being above the exceptionally low bar set by other businesses in terms of security. Ransomware groups are well-resourced and technically skilled, but as with any other profit-minded organisation they are eager to extract maximum profit for minimum effort. Closing down the easy holes, carrying out just basic security hygiene, requires them to expend more effort and so they will often simply move on to another victim. Until the entire world has got up to a basic level of security, this tactic will keep working.

The second part is to ensure thorough backups and a disaster recovery strategy is in place and tested. While ransomware that will lie low and hibernate exist, staying under the radar to infect backups, this is rare as again it requires additional effort and thought. Knowing that you can restore operations within a few hours after losing systems removes the threat of ransomware.

14.5   How about insurance?

AXA recently announced[2] that they would no longer be making payouts for ransomware part of their cyber insurance policies. Others may follow suit, as ransomware is one area where insurers

---

[2] Frank Bajak, *Insurer AXA to Stop Paying for Ransomware Crime Payments in France*

seem to have misjudged the risk. Worse, there are known cases of companies with insurance policies not only being targeted, but the ransom payments being set at a level known to be covered by the insurance policy. In effect, insurance companies have been subsidising the ransomware industry, driving up payments by making payouts available to their affected customers.

Security and Technical Debt Collection

Issue 60: 2021-08-17

You cannot work in close proximity to technical people, particularly those who build systems, for long without hearing the term technical debt bandied around.

Technical debt is what you are adding to every time you choose an easy or quick solution now, rather than looking at longer term strategies. It is the technical expression of 'failing to plan is planning to fail', and it has consequences.

If a technical system is held together with metaphorical prayers and duct tape, it will often be too fragile to effectively maintain (i.e. trying to update or patch it is likely to cause an outage, or just break it irreparably). Given the earliest systems in an organisation are also often the most important to its operation, the most critical systems are usually the ones carrying the most technical debt.

While it's often thought of as just adding friction to systems, possibly the occasional outage, it is much more insidious and damaging when we consider that most technical debt also involves what I like to call security debt. If you aren't able to update a critical system, to maintain it effectively, it will be vulnerable to cyber security threats. Known vulnerabilities are a key factor in most cyber security incidents, and these exist because systems go unmaintained.

This is a big enough problem on its own, but then we look at the world of Operational Technology (OT) and Industrial Control System (ICS) and things get worse, quickly. OT refers to technologies used to manage physical systems, often but not always industrial (building HVAC, access control systems, lifts, etc are also OT systems). ICS is a subset, specifically the systems that monitor, manage, and control industrial processes.

15.1   OT, ICS, and CNI

The most important processes involving OT are those in Critical National Infrastructure (CNI), everything from power plants to water treatment facilities. These systems were often automated before security was a major concern in the way it is now, connected to the internet to allow remote or centralised monitoring, and carry enough technical debt that they are often not even possible to maintain. Manufacturers for some of these systems no longer even exist, and the expense of them means that they are certainly not refreshed every few years as other IT systems should be. The first known, and most famous, attack against ICS systems involved the Stuxnet malware, uncovered in 2010. Stuxnet is still one of the most sophisticated cyberweapons developed to date, and has been repurposed to carry out other attacks after the one that (rightly) made it famous.

To keep the story short the Stuxnet malware was developed to compromise Microsoft Windows machines to gain an initial foothold on a network, after which it would seek out the controllers which automated gas centrifuges for separating nuclear material. Estimates are that Stuxnet ruined roughly one-fifth of Iran's nuclear centrifuges, and set back the national nuclear programme several years.

Stuxnet was developed to be subtle, it did not simply cause centrifuges to fail but introduced random variances in their operations which caused them to fail faster. It's estimated that it was a year after release before it was discovered, and the discovery was more luck than planning.

It isn't only malware that we need to worry about affecting CNI systems. While the Colonial pipeline attack did not involve any targeted ICS systems (it was down to a compromised password, and the shutdown was precautionary) if the attackers had taken aim at truly causing chaos rather than deploying off-the-shelf ransomware the attack would not have been detected as quickly, and potentially would have affected the pipeline's ICS systems. Given the control those systems had, the damage could have been much more significant. The attackers showed no signs of attempting to breach those critical systems.

In another recent case, an attack against a Florida water treatment plant was down to old software being in use (that technical debt again), and while it was detected and further safety measures were in place, the attackers adjusted the amount of sodium hydroxide (used in small quantities to lower the acidity of water, and in large concentrations capable of causing chemical burns) upwards about 100 times. Fortunately it was detected, but in this case we are seeing attackers deliberately and maliciously trying to cause damage — whether just to send a message and knowing that other measures would prevent the attack, or genuinely attempting to poison the water supply is unknown and only limited information is being shared.

So far there are only two attacks confirmed to have caused destruction of equipment (though a recent hospital ransomware attack has been held directly responsible for a loss of life). The Stuxnet attack damaged nuclear processing centrifuges in a very careful way. A second attack, some years later, occurred in Germany where a steel mill was compromised. The attackers managed to disrupt the control systems of a blast furnace to enough of a degree that it resulted in 'massive' damage, believed to be through overheating and removing the ability for the furnace to be shut down.

More seriously, or at least more impactfully, shortly afterwards the Ukraine power grid was deliberately targeted with a strain of malware named 'Black Energy', resulting in over 200,000 customers losing power. A year later, the same happened with a different attack using more sophisticated malware known as CrashOverride. Both of these were perfectly capable of being much more serious, as the attackers chose to only cut the power and not to reconnect it out of phase, which would have been catastrophic.

Another attack in 2017 was the first deliberately targeted at safety systems, designed to enforce emergency shutdowns when human life is at risk. The attack was originally believed to be a malfunction in the equipment, until the security team sent in to investigate determined it was due to malware and was part of an effort to develop the capacity to cause physical harm[1].

Realistically there aren't any easy answers to this technical debt problem. The effort and expense that would be needed to update these systems are beyond what the organisations who own them are willing to or able to afford, and the alternative solution of fully isolating the facilities requires re-engineering processes that are highly dependent on interconnection and effective communications.

It is vitally important whenever designing a new system to carefully consider the technical decisions being taken. The risks of not doing so are becoming more visible every day.

---

[1] *Triton Cyber Attack: Hackers Target the Safety Systems of Industrial Plants | SCOR*

Deep Fakes and Impersonation

Issue 61: 2021-11-07

Your principal pops up on video chat, asking you to change a route to go past a particular location, as their plans have changed and they have a meeting. Luckily when you make the pickup, you confirm with them. They're surprised at the request as they don't recall calling, and so you don't make the route change. Later investigation reveals that the call was made from a spoofed phone, using a real-time generated video and audio stream of the principal mapped to the live movements of some unknown adversary. It sounds like science fiction, or rubber masks from Mission Impossible, but the technology to do this is now so commoditized that you've probably seen it being posted on social media to let people duet with celebrities or insert themselves into film clips.

16.1   The Fake Obama

Possibly the most famous and dramatic demonstration was a video of President Obama seated in a briefing room. The opening lines are the most dramatic: "We are entering an age in which our enemies can make it look like anyone is saying anything at any point in time." This video was created in 2018, by Jordan Peele. The synthesized voice and image were synchronized to Jordan's own lip movements and words, and were generated based on publicly available video and audio of Obama speaking. I highly recommend searching for and watching this video if you want a Halloween scare.

Figure 16.1: Comedian Jordan Peele created a deep fake video of Obama.

Since then the technology has improved by leaps and bounds, and can now be done in near-real-time. In 2019 a fraud was carried out against a UK-based energy firm (still unknown which), where the CEO was convinced he was on the phone with his boss, the CEO of the parent company. He followed orders to transfer €220,000 to the bank account of a Hungarian supplier[1].

16.2   Is it a real threat?

This attack was purely voice-based, but the CEO stated that he specifically recognized the very subtle German accent in his boss' voice, and that the patterns and rhythms of his speech were the same. The fraudster carrying out the attack made three phone calls – first to initiate the transfer, second to claim it had been reimbursed, and once more to ask for a follow-up payment to be transferred. It was only after the second call, noticing that the supposed reimbursement had not happened, that the CEO looked harder and noticed that the call had been made from an Austrian phone number.

What is slightly shocking (and fortunate for the CEO) is that the fraudsters had not spoofed the mobile phone number to be from the right region, which is a trivial feat. If they had, the follow-up payment may have happened.

In 2020 a bank manager in Hong Kong received a phone call from a company director he had spoken with before, talking about an acquisition he was looking to make and that he needed some transfers (around $35 million) authorized. He recognized the voice, spoke with the nominated lawyer after checking his inbox for e-mails from the right sources, and began making the

---

[1] Catherine Stupp, *Fraudsters Used AI to Mimic CEO's Voice in Unusual Cybercrime Case - WSJ*

transfers[2].

These are the attacks which are known to be successful and have been made public. Other attempts have been reported using both video and audio, and they're unlikely to be going away any time soon. Detection technologies exist, but are not commonly used and the deception technologies continuously improve. We are likely to see a lot more in the future given the reliance over the last two years in online meeting tools and the opportunities for attackers to use them to manipulate circumstances in their favour.

Even ignoring the more attention-grabbing financial attacks, there is another attack vector available based on images. In 2017, deep fake pornography became prominent, with an eventual estimate in 2019 that 96% of all deep fakes were pornographic[3]. Celebrities were the most common victims, with some of the fakes featuring in articles, but anyone could be a victim.

Of course, if pornography is an option, blackmail materials indistinguishable from genuine images can be generated. Arguably there's the potential for this to then devalue blackmail material since anything can be argued as a fake. Deep fakes have also been used in politics (notably in 2020 by the Belgian branch of Extinction Rebellion who published a deep fake of the Prime Minister talking about a link between deforestation and COVID-19[4]), art (debates are ongoing about artificially inserting deceased, or even simply unavailable celebrities into media), fraud,

---

[2] Thomas Brewster, *Fraudsters Cloned Company Director's Voice In $35 Million Heist, Police Find*

[3] Ryan Duffy, *New Research Finds 96% of Deepfakes Are Pornographic*

[4] Gabriela Galindo, *XR Belgium Posts Deepfake of Belgian Premier Linking Covid-19 with Climate Crisis*

and to create fake social media profiles for non-existent persons for misinformation campaigns.

Figure 16.2: A face generated from thispersondoesnotexist.com

These fake social media profiles are often referred to as sock puppets, with the most well-documented occurring in 2018 when a persona named Oliver Taylor submitted articles (which were then published) accusing a British academic and his wife of being terrorist sympathizers[5]. While it is believed, given the evidence, that the persona is entirely synthesized, several newspapers have not retracted the articles or removed them from their websites.

16.3   What can you do about it?

- Verify – if a phone call is asking you to do something, confirm through some other means. Preferably through

---

[5] *Deepfake Technology Used to Make Journalist Oliver Taylor | Futurism*

a secure channel or in person. If an e-mail is asking for something, verify it. And, if the impersonated person or the target of the impersonation has the type of profile that attracts capable adversaries, do not take media of them at face value.

- You can test your own abilities to spot fakes with a tool created by MIT at https://detectfakes.media.mit.edu. Other similar online quiz-type tools exist, some more challenging than others. The site https://thispersondoesnotexist.com generates random synthesized faces with each visit, highly popular for the simpler form of sock puppet profile, and with some giveaways in the generated pictures that are useful for detection.

Ultimately though we are simply entering an age where the integrity of electronic media cannot be assumed, and need to adjust our behaviours and processes appropriately until the tools to protect ourselves become available (if ever).

Part IV

2022-2024

Learning Cyber

Issue 62: 2022-02-21

We're now into my fourth year regularly writing for Circuit, and while the cyber security scene has changed there are some obstacles which seem insurmountable. One of those is the often-discussed and disputed skills gap, the shortage of people in cyber security.

There's a lot of debate over the skills gap and I've mentioned it before.

For now I'm going to set that to one side, and talk about the various ways organisations and individuals try to address it, from self-study, to degrees, to boot camps.

17.1 Self-Study

Probably the most championed and well-known way to break into cyber security is self-study and self development, with something of an old-fashioned and outdated view that the best hackers are self-taught for free and so the best cyber security people must be as well. Dr Richard Diston, writes the "worst advice is the stuff you get for free because nobody has any investment in its quality". This is true in any domain, and especially in cyber security a lot of the free advice out there is put out for ego purposes as much as educational. While there is definitely some good material available, being able to filter out the bad stuff requires a degree of experience.

Even with paid self-study material, there's an investment of time and self-discipline required which works for some, and not for others. Equally there's cheap material which is, to be blunt, outright misleading (for that matter there's some expensive material which is actively harmful if implemented). Autodidactism is definitely an option, just one that requires a significant amount of work and can involve more than a few dead ends (and drops) on the way. If you decide to take this route, it is definitely worth joining some of the many communities out there for cyber security learners as they can help validate, assess, and guide towards the useful materials and filter out bad advice.

I am far from against the self-study route, but it carries risks that learners should be aware of and can be both frustrating and ineffective. Very much a case of buyer beware.

> ■ There is something of a cult of self-reliance in the cyber security industry, along with an unhealthy dose of anti-intellectualism. There are plenty of valid criticisms of all paths to learning, but none of them should lead to an automatic dismissal of that path. For the right person, any and all of them can have value and telling people not to explore one is doing them a disservice.
>
> Although I'll note that I'm talking about the routes to learning in general here - specific self-study materials, academic institutions, and boot camps are absolutely not above criticism and warning people away from. When the dust settles I have some stories to tell about fake universities, charlatan influencers, and bottom-feeding boot camps, but I've discovered an allergy to threats of lawsuits so I'm trying to avoid them.

17.2   Certifications

Another route, also often touted as the one true path in, is to take various industry certifications either through a self-study route again (just with expensive course books, or less expensive unofficial course books), online courses studying over the long term, or short boot camps promising certification at the end.

Certifications can definitely help. A lot of the time entry level roles are poorly specified by potential employers with a raft of required certificates, and having a few of the right ones can help get past the human resources filter to speak with an actual human being who can evaluate knowledge and understanding effectively. Many (though far from all) certifications also have reasonably good, well-researched source materials which effectively teach vital areas of cyber security for new entrants.

Sadly, very few of those certification courses assess that knowledge in a way that makes the certification useful. The vast majority rely on multiple choice automated quizzes, and seem to believe packing them full of a family edition Trivial Pursuit game's worth of occasionally-tricky questions is the way to assess learning.

The few that don't are easily recognisable as they tend to take a more hands-on approach to assessment, offering a variety of laboratory tests and simulated environments to demonstrate learned skills. These work well for the better-defined, more technical roles within cyber security such as penetration testing (trying to exploit systems to gain unauthorised entry or cause harm, with permission from the owner of the system) and Security Operations Centre (SOC) analysis (monitoring, reacting to, and trying to prevent the same harm). SOC analysis work is not as well defined, since the roles can encompass anything from straightforward monitoring through to aspects such as forensics and vulnerability management, but a good course will take you through these, and a good lab assessment will have you carrying them out.

The problem is that very few of these practically-assessed exams are well recognised so far. The OSCP is the most well-known for penetration testing, though others have started to pick up similar approaches. For SOC analysis a relative newcomer which is gaining ground fast is BTL1. OSCP is now often found on penetration testing roles, while BTL1 has more awareness work to do before it becomes commonplace. Of course when you do see a job description with these certifications, it's a good sign that the hiring manager understands what they are looking for in a qualification.

17.3   Boot Camps

There are various boot camp programmes out there, and for full disclosure I teach part time for one called CAPSLOCK. As with anything else quality varies, so I would always say to try and speak to previous students of the course and get their opinions. Many of them are, sadly, focused on teaching a handful of certifications and promising ridiculous salaries which never materialise. Fortunately, a growing number of others are more interested in helping to resolve the skills shortage.

Boot camps get a mixed bag of opinions in the industry, many people are very attached to the self-taught route and accuse these courses of profiteering and exploitation (and indeed, some do exactly this), without any evaluation of the individual courses. Others will take a dislike to even the best course because it does not include the certification they felt is most important (which is frankly often a sign of a narrow view of the field).

The good boot camp courses will include placement support and advice to get learners not only through the course, but into an appropriate role, which brings me to the last piece of this article.

17.4　Breaking into cyber security

> ■ I really dislike the whole 'break into' concept and terminology, especially the way it's presented as the only route in. Whenever I'm talking to people trying to start or pivot their career path in I tend to talk about sneaking through the service door instead.

Over the years I have helped a number of people to enter the cyber security field, and the thing that I have found most in common among the ones who succeed is that they did not get their roles through simply applying to jobs online. A few have found good recruiters that they have built up relationships with over time, but most have relied on networking and making themselves known to people who are likely to be looking for recruits.

If you are interested in entering a cyber security role, my strongest piece of advice (beyond learning whatever aspect or aspects interest you) is to put yourself out there. That's not a case of simply sending your CV to as many places as possible, but includes speaking at rookie (or non-rookie if you are comfortable with public speaking) conferences, networking, using LinkedIn, joining communities, and so on.

I will not pretend that I believe entering cyber security should require these things, but there is so much misunderstanding in the sector, and so few of us out there (DCMS estimated 46,000 cyber security professionals in the UK in total, 30,000 of them work for large vendors and include sales staff while the rest work for organisations directly), that it's the system we have until we manage to change the way the recruitment works.

> ■ The skills gap is a topic of much debate. The two loudest claims are that it does or doesn't exist, but the more nuanced view tends to be that it simply isn't where people believe. Rather than being a problem of not enough people wanting to enter the industry, we lack the support network for those who do to move to the next level. Entering a field of any kind requires time and support to develop competence, and the distribution of people in the cyber security industry doesn't provide for this support.

Building on Quicksand

Issue 63: 2022-05-29

The vast majority of modern internet technology is built on foundations that make the average sandcastle look earthquake-proofed.

It's a fundamental flaw in the way that software engineers and developers operate and interact with the open-source community, combined with the massive complexity of code that exists today. Those who aren't involved in the development of software rarely understand how many layers upon layers of complexity it's built upon - in fact many of those who are involved in software development fail to understand quite how unsteady and teetering their work is.

There are two fundamental problems.

18.1 The Problem of Standards

One is that the requirement for all modern software and hardware to interoperate effectively with any chance of working (which, given the number of different systems involved is a miracle in itself) means everything is working with a standard set of protocols for communication and messaging. These have been tried and tested over the years, but every now and then an unexpected problem comes up with them that means everything breaks.

In 2014, on an otherwise normal Tuesday morning in August, large parts of the internet simply stopped responding or only responded intermittently when people tried to contact them. This wasn't due to a sophisticated attack, or damage to an undersea cable, or even a fire in some data centre. Instead it was a problem with memory limits and Border Gateway Protocol (BGP), the protocol which allows separate networks connected to the internet to communicate effectively with each other.

BGP uses a set of routing tables to route messages and in some older hardware, critical to the functioning of the whole system, the memory chips used meant that it was limited to 512,000 entries. When the tables started to reach that and exceed it, they started to fall over. The extent of the damage isn't well recorded, but it covered major websites and was felt worldwide for multiple hours.

Like BGP, many of the protocols our technology relies on were developed before our systems reached the complexity and interconnectedness they have today. While some have been redeveloped, redesigned, and rewritten to account for changes and provide future-proofing, getting these new protocols rolled out raises its own challenges (if you ever want to upset a network engineer, ask them how the IPv6 rollout is going and whether they'll be ready to decommission IPv4 next year).

Even where the protocols themselves don't throw up fundamental flaws when new circumstances arise, software developed to use the protocols can have its own problems. Also in 2014, this time in April, a vulnerability nicknamed Heartbleed was revealed in a piece of software called OpenSSL (it had been introduced into the library in 2012). This flaw meant that anyone with fairly basic technical knowledge could query a supposedly secure web server running the library and retrieve random sections of memory - which could contain anything from random useless data to unencrypted passwords.

A patch was released the same day, but even a month later it was estimated that 1.5% of the top 800,000 websites using the protocol that OpenSSL enabled were still vulnerable. Even in 2019, a tool called Shodan reported 91,000 vulnerable devices were still online. With these shared protocols we have risks of obsolescence, discovery of vulnerabilities in the protocol themselves (WiFi encryption used to use a protocol called WEP, now breakable to retrieve the password for a network and decrypt all traffic passing through it with a few minutes' work, a laptop, and a cheap WiFi dongle). We also have problems of vulnerabilities being deliberately or accidentally introduced into the software libraries which provide these protocols and are integrated into applications across the planet (often with no real maintenance effort).

18.2　It Gets Worse - Jenga!

As if these problems were not bad enough we face a whole new class of issues when we look to less-essential but more useful software. Many applications integrate standard, open-source libraries to provide key functions. Open-source software has its code published online under varying licences for anyone to use as part of other projects.

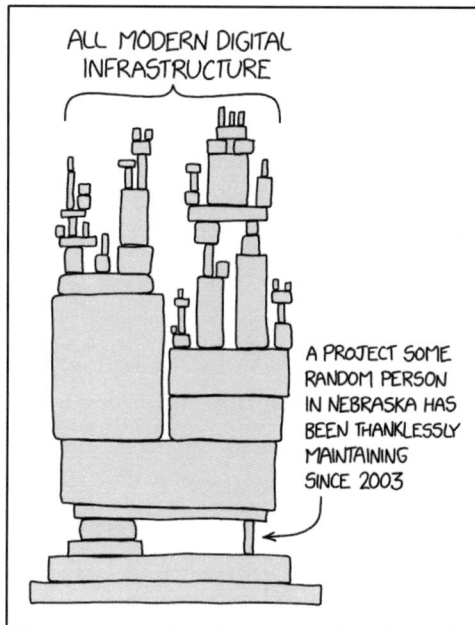

Figure 18.1: Dependencies by Randall Munroe of https://xkcd. comxkcd, reprinted here with permission under Creative Commons Attribution-NonCommercial 2.5 License.

In March 2017 around the world applications suddenly broke

with no warning, to the despair of developers everywhere[1].

A man called Azer Koçulu had received threats of legal action from a trademark lawyer representing a company called Kik. The story rumbled on for a while, and then the software repository where the package Kik objected to took action and renamed it, ignoring any protects by Koçulu.

Objecting to this action Koçulu, who had written over 270 packages and hosted them on the site for anyone to use, decided to take everything and go elsewhere.

One particular package was only eleven lines long. It did very little, just a simple utility function, one which companies had built into their software in a fundamental way often without their knowledge. Companies such as Facebook, and ironically Kik themselves. That package had been downloaded about two and a half million times in the previous month, and now every single one of those complex applications catastrophically broke.

Despite this important lesson, little has changed today with software companies still using open-source components and trusting them as part of ever-more complicated applications. I've written before about supply chain threats, but even where organisations are acting to limit threats from managed service providers and application vendors they rarely understand the fragility of the underlying structure and there are no signs of things getting better soon.

To put it bluntly, the shining edifice of modern communication technology, no matter how impressive it may be to look at, is

---

[1] Keith Collins, *How One Programmer Broke the Internet by Deleting a Tiny Piece of Code*

held together at its foundation with little more than spit and chewing gum. Failures are unpredictable and often dependent on the goodwill (or bad intentions, or grudges, or inattention) of those who volunteer their time to support the software on which it all rests.

This year alone we have seen several critical and widespread vulnerabilities affecting thousands of systems, due to flaws in open-source packages unsupported by the companies who make use of them. It's likely these will only increase as more people look for them, so make sure any technology you work with is kept up to date with security patches from an authorised source and keep an eye on what's used in-house.

> ■ This...has not improved since it was written. If anything it's just worsening with the addition of generative AI creating code which is neither well-documented, nor understood. The critical dependency web is growing.

Can AI Impersonate a Human?

Issue 64: 2022-08-24

Given a recent bit of controversy around Google's new AI, LaMDA, being claimed to be sentient[1] I thought it was a good time to talk about AI again. If you've not come across it, an engineer from Google (now suspended) claimed that he was convinced the AI was 'sentient'. In judging these claims, the chat log he used as 'proof' was heavily edited for readability, and was compiled from nine different conversations with the chatbot, so conscious, sentient AI is still a long way off.

I've spoken about Deep Fakes and AI-generated content before, but while there have been a few incidents of faked voice calls for scams, and faked videos for humour, Russia's activities in Ukraine have highlighted how these can be weaponized for propaganda purposes.

In May, the Ukrainian government released a series of audio recordings in which Russian officials were allegedly discussing

---

[1] Khari Johnson, *LaMDA and the Sentient AI Trap* | *WIRED*

the situation in Ukraine. The twist? The people speaking on the recordings had been replaced by AI-generated voice clones.

This is far from the first time that AI has been used to create fake content. In 2017, a company called Lyrebird released a platform that allows anyone to create an AI-generated voice clone from just a minute of audio. The results are impressively realistic.

> ■ In 2023, I even ran an experiment with the BBC's Morning Live team to impersonate one of their presenters over the phone. It worked...briefly...before the target asked whether they were putting on an American accent. Not bad for a few minutes work putting it together though.

In the wrong hands, AI-generated content is used to create fake news stories, spread misinformation and propaganda, and even to generate fake reviews and testimonials.

As AI gets better at generating realistic content, it's becoming more difficult to tell what's real and what's fake. This poses a serious threat to our ability to distinguish between reliable information and disinformation.

Deep fakes are a particular concern because they can be used to create realistic, believable audio and video content that is very difficult to distinguish from the real thing. This means that deep fakes could be used to spread misinformation or even to impersonate people for fraudulent purposes.

19.1 Can you tell the difference between AI and human content?

Now this is a bit of an experimental post, and I'll explain why in a bit, but I want to ask how sure you are that you can tell the difference between something created by an AI and something created by a human.

To help with this, I've got two pieces of content. One is AI-generated, and one is human-generated. Can you tell which is which?

Figure 19.1: One is a photograph of a dog on a beach from Meg Sanchez on Unsplash, and the other is a completely artificially generated picture of a dog on a beach from an AI tool called DALL-E 2. To avoid spoiling the surprise, I'll tell you which is which at the end of this article.

All fairly light-hearted so far, but this has real consequences for security on individual and national levels. Let's try another one.

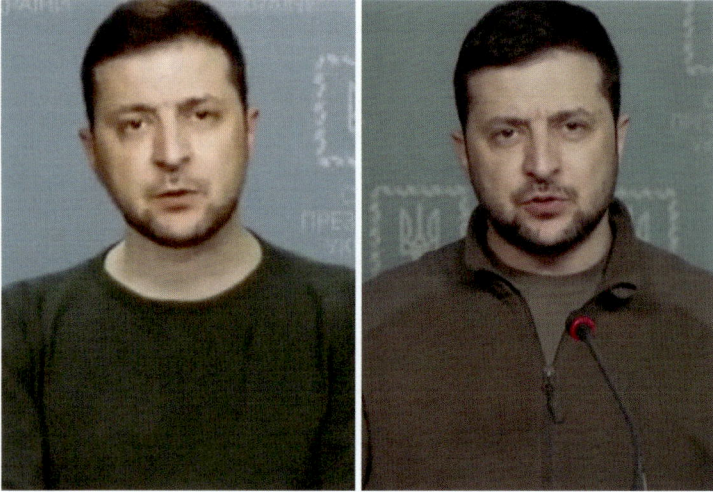

Figure 19.2: One of the two pictures above is a frame from a genuine video. The other is taken from a faked video of Prime Minister Zelenskyy announcing peace and demanding Ukrainian troops lay down their arms. The video was established as a fake quickly, and generally ridiculed for poor quality, but there are much more sophisticated fakes out there.

We're very much past the time of Deep Fake technology being used to create fake pornography for sale or for extortion purposes, which has been going on for years. With technologies like DALL-E 2, 'proof' of guilt can be created from nothing more than a description in photographic form, and it won't be long before that applies to video as well. Deep Faked audio is already sophisticated enough to be difficult to distinguish from reality, as you can find out if you try Lyrebird.

Of course, it doesn't stop there.

19.2 Can you tell the difference be-tween AI and human writing?

There's a second experiment at play in this article. Roughly half of it is written by an AI system, while the other half is written by hand (well, typewriter). See if you can tell which is which.

> ■ A slight problem here is that reading it back now, I didn't actually mark which parts were written by AI and which were written by me.

The reason why I'm doing this experiment is to show just how good AI systems have become at imitating human communication. In particular, I wanted to test whether an AI system could successfully impersonate a human in any context.

So far, the results have been mixed. On the one hand, the AI system has been able to fool some of you into thinking that it is human. On the other hand, other people have been able to tell that it is not human. Of course, that does mean some of you have decided that I'm not human either (unless that was written by the AI of course, this could get a bit confusing).

Regardless of the outcome of this experiment, one thing is certain: Deep Fake technologies are getting better and better, and are a genuine threat on both individual and societal levels through misinformation, potential extortion, impersonation, and myriad other attack vectors that we haven't even conceived of yet. Tools to recognize generated content are far less developed and less common than tools to generate it, and so currently maintaining suspicion of anything that seems out of character and validating content through other means is the only approach that seems viable.

Incidentally, with both of the sets of pictures above, it was the left-hand one which was the fake.

Close Protection in Virtual Worlds

Issue 65: 2023-01-04

Over the course of the pandemic and various lockdowns world-wide, many people were forced online to interact in ways they hadn't before. Platforms like Zoom provided video chat, but for many people, these very cut-down conference calls provided only a minimal substitute for genuine interaction with other humans. As a result, over the last few years, the number of virtual worlds where participants can interact more realistically and with more autonomy has grown, and rapidly.

One of the earliest examples that still persists today is Second Life, a 3D virtual world originally released in 2003 and still going strong today with its own economy, communities, and hundreds of thousands of virtual users. It would take nearly two decades before things progressed to virtual reality headsets being commonplace enough that now we have people more directly interacting, but even back in the early days of these virtual worlds (or the metaverse if you want to sign up to the latest buzzword to describe them) there were signs that these virtual worlds were far from comfortable for high-profile participants.

In 2006, three years after Second Life launched and shortly after the first virtual millionaires emerged (real money, from selling virtual real estate within the world) an interview with one of these high net worth individuals, Anshe Chung (the name of their avatar, with the real name Ailin Graef) was disrupted by a horde of animated flying penises for fifteen minutes. Retreating from the assault, Chung and their interviewer retreated to Chung's theatre, where the assault continued and eventually caused the server to crash[1].

At the time, this was seen as a bit of almost harmless trolling by many in communities aware of these worlds, but as more and more people enter them and spend time in them, these have taken on a new perspective.

I have a side business running virtual events, and one of the things we find more and more necessary as companies start latching on to the possibilities of these platforms is the need to provide some sort of protection for high-profile individuals taking part. Recently, virtual close protection officers were required for a large virtual event where executives were present, to prevent them from being mobbed in the virtual world by attendees complaining, questioning whether they were really speaking to the CEO, or otherwise intruding in personal space in a way that they would hesitate to in reality.

This all takes place in a virtual world more akin to Zoom, with only 2-dimensional avatars wandering around, rather than anything more sophisticated, but it's still a serious issue for those

---

[1] Ross Miller, *Second Life Millionaire Pummeled with Penises*

high-profile participants if they want to engage with employees or the public without facing harassment. While the threats may not be of the same severity, if their security is only considered in the physical world when they are exposed to others in virtual worlds there are still threats that will apply. Even in some virtual worlds, as odd as it may sound, dangers of eavesdropping and intruders in private conversations can become an issue once again.

These issues only increase when we consider true virtual reality worlds where instead of representation by an abstract avatar on a screen, those taking part feel like they are directly immersed. We are still some way from these being a full substitute for reality, but for those who have never experienced today's virtual reality systems, even the most cartoony or unrealistic forms can be surprisingly immersive and engaging - and with that immersiveness and engagement come threats to people's safety. Plenty of research has shown that even in the less immersive, two-dimensional forms, many people identify very strongly with their avatar to the point of reacting to attacks or interactions with them, and this only strengthens when we are taking part in a three-dimensional world from a first-person point of view[2].

As yet there have been few incidents in modern virtual reality to compare with the 2006 flying penis hordes, but only because this is still an emerging technology and high-profile individuals have only taken limited part in it and only in carefully controlled conditions. Earlier this year, shortly after Facebook's announcement of their Meta VR platform, we had the first reports of harassment and sexual assault against a researcher[3]. This was

---

[2] Wolfendale, "My Avatar, My Self"

[3] Tanya Basu, *The Metaverse Has a Groping Problem Already*

in a public, open chat room, and while there is an argument that of course the targeted user could have left the environment unlike in reality there are issues which embolden harassers and attackers behind a veneer of anonymity and safety from any physical consequences.

> ■ Sadly it is no longer true that there have been few similar incidents, with cases on the rise and including what has been described as a gang rape of a minor in virtual reality which is under investigation by police at the time of writing[4]. As the technology becomes more popular and accessible, these incidents will only become more numerous.

We are still very much at the early days of these technologies and it is possible that we will see providers bringing in controls and tools to protect users (though for those familiar with the history of social media and moderation tools on the internet, extremely unlikely). Most likely though all we will see is use of these tools increasing over the next few years and more of these issues coming to light, especially if or when this form of virtual reality becomes a place for high profile individuals to engage with others. At this point there is going to be a need for people who understand the technology, the cultures of these virtual worlds, and how to de-escalate and deal with situations that arise within them.

---

[4] Sales, "A Girl Was Allegedly Raped in the Metaverse. Is This the Beginning of a Dark New Future?"

FUD for Thought: AI Armageddon and Security

Issue 66: 2023-05-04

Fear, Uncertainty, and Doubt (FUD) is the dark reflection of hype. Given all the marketing and tech utopian hype about AI, the amount of FUD going around these days is unsurprisingly high.

I'm not a futurist (studies show that they're often less accurate than flipping a coin in any case, despite their own claims) so it's worth taking any predictions I make here with a good handful of salt. What I am going to do is try to cut through the hype, talk about where the technology really is, and why while it may be attracting a lot of investment it's not the world-changing threat that it gets presented as.

> ■ I have issues with the futurist industry, since their predictions work in roughly the same way as Nostradamus' and are statistically speaking about as accurate.

Let's start with the big one, ChatGPT and Large Language Models. These tools have been around for a while now, longer than they've been hitting the headlines, and I even used one before ChatGPT was launched to part-write an article and challenge people to spot what was human-written and what was machine.

21.1   Manufactured Mediocrity and Chat-GPT

All of these generative models, whether language or imagery such as Midjourney, make for some impressive gimmickry. There's a writer's strike on over the use of them in media, there's fears of them replacing writers, artists, and creators of all types. All of these are not entirely unfounded – AI-generated content is multiplying by leaps and bounds every day.

The biggest challenge it has isn't that these generative models don't truly create anything new (although they don't, more a lowest common denominator of the input fed to them), or that they're currently suffering constant degradation from ingesting more and more AI-created content (which leads to a beautifully named condition called model collapse). It's that they have no understanding of their output.

Take ChatGPT – it's great at creating reams of very plausible-sounding text which supports whatever it's been asked to create. Already we've had lawyers reprimanded for relying on non-existent legal precedent because they relied on AI to do their research. Those reams of text are also, well, utterly mediocre.

So what does this mean for the security industry?

Not much, honestly. Generative models create bargain-basement content and they are likely to devalue skilled content creation – you can look at the translation industry for parallels as machine translation has been around for decades. Yet human translation still hangs on, because people who care about their content have seen what happens with poor-quality machine translations without at least some error-checking by skilled individuals.

Expect a lot of security awareness content being machine-generated to appear. Along with a lot more plausible-sounding misinformation and disinformation. We're all going to have to get better at critically evaluating the media we ingest, but that's not a new problem. Ultimately though, for most of the industry generating content isn't the bread and butter – and that's all that generative models do despite claims of being able to solve all the world's ills or cause them.

21.2   What About Decision-Making?

Different AI models are used for different purposes, and more likely to have an impact are the decision-making Artificial Intelligences. The best example of this, and the flaws in it, is AI being used to evaluate risk in parole board hearings in the US.

I've covered this before so I'll only touch on it lightly here. Without very careful evaluation, and dealing with biases in the data, these risk evaluation systems not only suffer from exactly the same problems and biases as humans – but can even reinforce them.

Worse, the models can't be questioned to explain their reasoning. They're a black box into which you can pour data, and get an answer, with no idea of how one led to the other. Humans here have a huge advantage in any risk evaluation – they can be held accountable for mistakes, they can explain their reasoning, and they are more likely to be questioned on their decisions where AI is often implicitly trusted.

21.3   Anything Positive?

There is likely to be some disruption, as people place too much trust in technology and abdicate responsibility. There will also be companies trying desperately to replace people with AI. To sum this up, a common joke in technology circles is that a piece of code it would take a programmer a day to write can be written in seconds with AI. The same programmer then has to spend a week debugging and fixing it to get it to work. Hardly labour-saving.

Where AI can be used well is in assistance functions. Recognition algorithms are already used to highlight features of interest in CCTV surveillance, helping guide that all-important human assessment to places where it's effective. Environmental monitoring can be used to alert for dangers proactively, instead of waiting for someone to check sensors, reducing the burden on human operators so that they are able to spend effort where it's most useful.

Even in risk intelligence, AI can be useful (though maybe not ChatGPT) highlighting information of interest and worthy of attention for specific issues, as our feed of information grows ever more unmanageable. Again, this only works if that information is then fed to human analysts who can turn it into intelligence.

There are benefits to these tools, with the right human oversight and effort, and a solid consideration of ethics involved. In some areas they may change the world – in medicine, for example, generative models are great for development of new and more effective drugs – but in an area so fundamentally human as security they're unlikely to make a big difference any time soon.

A few years ago one of the first 'security' robots called Steve was being used to patrol a shopping centre. It missed a patch of loose surface, and promptly drove itself into a fountain (Figure 21.1). Others have been deployed more widely, thank to the manufacturer's heavy marketing, but after years there's little to indicate they've had any significant impact.

Figure 21.1: A step along the road to developing Marvin the Paranoid Android. Image credit to x.com/gregpinelo

Just like Tesla's accident-prone 'self-driving' cars, we shouldn't expect robots to be replacing human security professionals any time soon.

Cyber Security, Christmas Gnomes, and AI Disinformation

Issue 67: 2024-02-16

In September 2023, an article took the world by storm, talking about Christmas gnomes being deposited in people's gardens to test whether homes are empty or open to being burgled. Over the course of around 12 hours, I followed this story as it made its way from the Mirror headlines, to most of the mainstream UK media, Australia, India, the US, and Germany. It may have reached other countries as well[1].

---

[1] Kerry-Ann Mills, *Police Issue Sinister Warning to Homeowners after 'Christmas Gnome' Found in Garden - Mirror Online*

> ▰ When I was first contacted by a reported about this
> it was early on in the cycle and I was in a car being
> from Nova Gorica to Venice airport, from a conference.
> Over the course of the two hour drive, and the wait
> for the plane, I watched the news headline going from
> one mainstream newspaper to worldwide. It was dra-
> matically eye opening and the only thing missing was
> popcorn, as I just couldn't look away.

I even had a few phone calls come in asking me to pop in for interviews to talk about the gnomes. That was what got me looking into it more seriously.

Figure 22.1: The sinister Christmas gnome.

There was only one problem with the whole story, no one had bothered to fact check it. In fact, as far as I could tell as I was following its journey around the world, there was very little human interaction with the story at all.

I talk a lot about mis- and disinformation, and this is part of the reason why. This particular story was mostly harmless (apart from spreading fear of course), but the fact it spread so quickly, so widely, without anyone tracing back to the source should start alarm bells ringing. Loudly.

22.1   What Happened?

North Wales police had a report of suspicious activity by two men a few days before the story exploded. They were reported to be looking at people's homes, and throwing items into gardens – including one soft toy gnome.

There was even, briefly, a notice on the North Wales Police website about it. For about an hour before it was taken down.

Suspicious activity - Broughton

Resident

Good morning,

North Wales Police received a report of two males walking around in the area of Somerford Road, Broughton on Friday morning (1st September). The males were behaving in a suspicious manner, looking at properties and throwing/placing items into resident's front gardens (one such item being a small soft toy gnome). They were carrying red 'royal mail' style shoulder bags, and one of the males was said to have a Birmingham accent.

If anyone else has seen anything, experienced this themselves or can identify the male from the attached photo, we would advise you to contact police and come forward with any information that could assist our enquiries into this matter.

Figure 22.2: The police warning.

So how did this suddenly flare up into worldwide news?

As best I can tell, the first major story was on the Mirror, and came with a disclaimer which gives a big hint as to what happened.

Every subsequent article used the same picture, similar text and article structure, and was timestamped later. The Mirror article seems to have been the initial vector for mutation from quirky police warning to mainstream news story barely resembling the truth. There may or may not have been human editorial review, but if there was then it was cursory at best and did not involve any research deeper into the story. The gnome morphed from a single item from a bag into the main story, and even got itself described as a calling card for burglary teams.

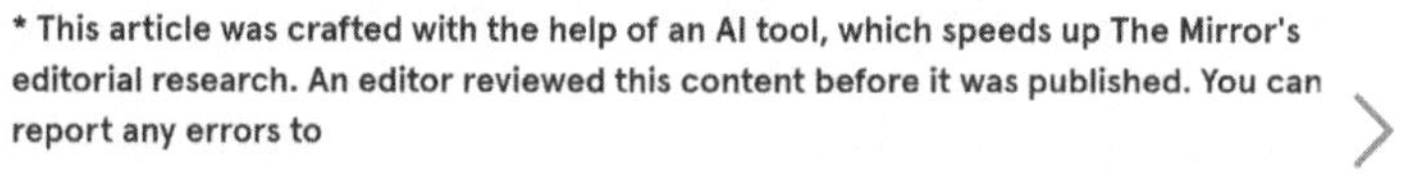

> *This article was crafted with the help of an AI tool, which speeds up The Mirror's editorial research. An editor reviewed this content before it was published. You can report any errors to

Figure 22.3: The disclaimer on the Mirror article.

From there, the similar language in every following article (including the foreign ones) suggest a lot of copying, likely using automated tools. Once the story was spread wide enough, it effectively became accepted as true, and made the leap to radio and television as well as going international.

22.2 What Does This Have To Do With Cyber Security?

During lockdown, stories and conspiracy theories circulated about dangers of 5G networks, culminating in physical attacks on telco engineers and arson attacks against the transmission masts. There is evidence that these stories were deliberately amplified by hostile nation-state actors, to sow chaos[2]. 5G masts are part of our critical national infrastructure. For the price of a few Facebook posts, a hostile power managed to radicalize domestic terrorists to attack critical national infrastructure during a worldwide crisis.

A few months later another fake Facebook event about a non-existent left-wing protest managed to gather a gang of 200 right-wing counter-protesters, largely heavily armed, to a location in the US. The same hostile power is generally believed to have been behind this.

> ▪ So far this use of misinformation to create and control a militia hasn't been repeated, at least as far as I am aware, but given current events around elections in the US I might not be surprised.

Fake news on social media is a big enough problem in its own right, but seeing how quickly we can have a story full of misinformation spread makes it even more alarming.

It isn't a stretch to imagine how this could be weaponized – and there are already signs of this having happened. Fake allegations about security flaws in companies, or in their software (or over-hyped allegations in some cases) have been used to manipulate stock prices. Equally, conspiracy theories about high-profile peo-

---

[2] Mark Lynas, *Anti-Vaxxers and Russia behind Viral 5G COVID Conspiracy Theory*

ple are a staple of modern-day social media culture and leave many of these people at high risk of attack.

Even spotting the dis- and misinformation isn't especially useful, as knowing it's false doesn't help to combat its rapid spread. Instead, high profile individuals should seriously consider investing in reputation monitoring and management, linked tightly to any physical protection they have to ensure safety. Those in close and executive protection need to be aware of the potential for viral misinformation to spread, and even more aware when it has potentially placing their principals at risk.

Part V

Unseen

Cyber: Drones, Smart Homes, and Social Media

Unpublished

While arguably cyber security has been around since the 1970s, it's only much more recently that the term has become commonplace. Even now there are regular debates about whether or not cyber and information security are interchangeable, and what exactly cyber security covers. These are unlikely to reach a conclusion any time soon.

Despite a lack of clear guidelines, it's important to recognize that cyber security is, like all other forms of security, ultimately about the protection of assets from threats, relying on principles of risk management. The only difference in cyber comes from the different attack vectors involved, many of which overlap heavily with other security domains.

Our technology systems are becoming ever more interconnected not only to one another but also in their interfaces with the physical human world. Some estimates suggest that we have over 46 billion connected devices[1], with the average household having more than 10 in 2020[2]. These devices include traditional ones such as computers, phones, and tablets, along with others such as children's toys, baby monitors, cars, lights, kitchen appliances, and even toothbrushes.

Along with the risks that arise from all this technology, cyber security also concerns itself with issues such as social media disinformation, ethics, and social attack vectors using technology. A whole area of cyber security focuses on social engineering, manipulation of people through various means to perform actions harmful to their own well-being or divulge information. Another, open source intelligence, looks to gather and analyse information passively about targets using only publicly available sources, and can have astounding results with the trust that individuals place in social media platforms.

23.1   It's About Systems

It is vital to remember with cyber security is that it's almost impossible to consider any aspect in complete isolation. While we simplify things as much as we can, the whole domain is about interconnected systems. Even when we begin talking about fully air gapped systems (ones not physically or wirelessly connected to any external networks), methods are constantly being developed to infiltrate, compromise, and exploit them whether remotely or through social engineering to deliver malicious pack-

---

[1] *'Internet of Things' Connected Devices to Triple by 2021, Reaching Over 46 Billion Units*

[2] *Topic*

ages within supposedly secure networks.

"The perimeter model is dead." – Bruce Schneier

Given the complexity of networks and the usefulness of connecting to them, a common view is that we cannot rely on traditional technical perimeter security, instead requiring defence-in-depth approaches tying together multiple diverse controls. We also have to accept that, even more so than when dealing purely with the physical world, no defence is perfect, and the threat landscape changes at such a pace that we must continually maintain, review, and improve our knowledge, processes, and technology.

23.2   Information and Cyber Security: A History

There is a significant overlap between the two domains of information and cyber security, with many people using them interchangeably. Both focus on a model known as the CIA triad, the confidentiality, integrity, and availability of systems, with information security also incorporating physical security concepts to protect information, while cyber security tends to focus more on the technical side of information security. Given the construction of our model information environments and their heavy reliance on technology, it is sometimes argued that cyber security makes up the majority of information security, and in many organizations, physical security is lumped into the cyber or information security department.

Truthfully the two approaches are so intertwined that it's unlikely there will ever be clear and comprehensive definitions, and the goals and principles are still fundamentally identical to the wider security discipline. Just as much as any other area of security, cyber security looks to protect assets from threats by applying controls. It is only that the potential interaction points between threats and assets in the world of cyber security are so numerous and diverse which makes it appear so complex.

23.3 Drones, Smart Homes, and Social Media

To understand why this diversity and complexity is important but does not fundamentally change the approaches we use with security controls, three superficially different networks are worth looking at.

When talking about drones we may be discussing highly advanced intelligent military drones or cheap, disposable, numerous drones using swarm technology to create impressive light displays. In either case, the devices themselves will be connected, even if only occasionally, to a network to issue commands.

While we should not completely ignore the different technologies involved, when we're considering cyber security at a high level we can reduce this to either a centralized communications model (i.e., a command post issuing messages to a number of highly advanced drones) or a mesh model (a swarm of drones relaying messages within their own communications network at need, issued from any point in the mesh).

A smart home, while it may seem completely different at first glance, is again simply a network of devices that can either be centrally controlled (often remotely controlled from a cloud interface), or in rarer cases follow a mesh model once again. Even with social media, ultimately it is a network of endpoints, though in this case the endpoints are people, who can send messages in various different ways through the network.

Ultimately all three of these and any other area of cyber security break down to the confidentiality, integrity, and availability of those messages in the network, and everything else descends from there. The aims of cyber security are to ensure that the message arriving at any particular node in the network is only accessible to those who should be able to receive it, has not been tampered with or spoofed by some unauthorized party, and is available to the node when it is needed. This applies just as much when the message is an instruction to a smart heating system to ignite the boiler, as when telling a drone to switch patrol patterns, as when a trusted friend on social media is apparently reaching out for help with a crisis.

The details of the technology vary, but if you can understand risk, control, and the ideas behind the CIA triad then you already know the fundamentals of cyber security and why it is important to incorporate it into any organization's understanding of security.

Self-Destructing Cars

Unpublished

In 2023, Tesla recalled more than 2 million cars over a defect in their driver assist software.

The defect was that it was too easy for drivers to misuse the system, which may be why Tesla also holds first place on the rankings for highest accident rate[1] in the US for 2023.

Self-driving cars have been the stuff of science fiction for years. If we want to be honest with ourselves, they still are. The technology is improving (ever so slowly), and they can run test courses happily enough, but dealing with the reality of other drivers on the road is a problem for them, and when you're nearly a tonne of metal you become a problem for everyone else.

Problems in San Francisco ran the gamut, from causing gridlock on residential roads as self-driving cars sat and waited for each other to make a move to serious accidents. If you're concerned

---

[1] LendingTree, *Ram, Tesla and Subaru Have the Worst Drivers*

about a self-driving car apocalypse however, take heart from discoveries made by residents that they could immobilise the cars by placing traffic cones on their bumpers - much in the way of superstitions about supernatural creatures and being unable to cross lines of salt. A vigilante group, called Safe Street Rebel, started roaming streets immobilising cars in this way in response to their city being used as a testing ground.

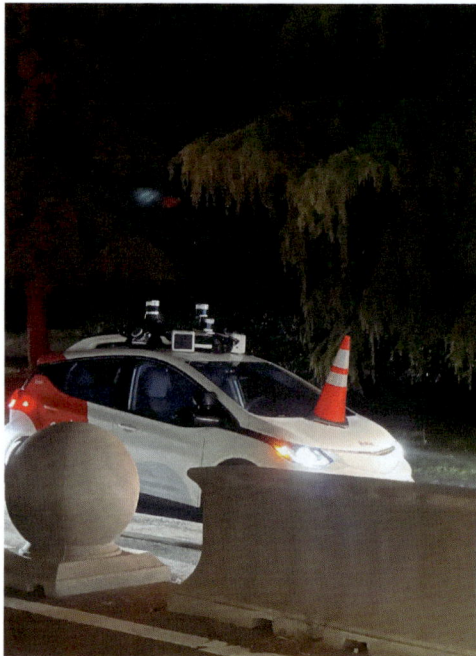

Figure 24.1: A wonder of modern technology, disabled by a traffic cone. Credit to Safe Street Rebel.

At any one time, a few hundred self driving cars were in testing. Over the course of the test they have racked up an impressive list of incidents[2]:

2 *ConeSF*

- Drove dangerously - straddling lanes, failing to signal, driving back and forth through red lights with passengers inside, and plenty more

- Went into an infinite-point turn cycle, driving back and forth endlessly

- Stopping in the middle of traffic, driveways, blocking petrol stations, and various other obstructions supposedly to pickup passengers

- Blocking drivers into parking spots

- Driving through emergency crash scenes and nearly striking first response personnel

- Driving into a ditch, again with passengers

- After a woman was hit by a human-driven car, ran her over and dragged her for ten metres at 7mph (the company hid the video from investigators, which led to their license being suspended)

- Blocked an ambulance carrying a critical patient who later died in hospital

- Ran over and killed a pet dog.

It's safe to say that self-driving cars won't be replacing people any time soon, despite the hype and billions of investment poured into their development. Which is almost certainly a good thing for already-crowded roads, as the volume of cars is already magnified by single-passenger cars and adding zero-passenger cars to the mix would not help.

The risks are obvious. Self-driving cars might be all the rage, but for anyone arranging security for a principal (or indeed themselves) not only is it definitely best to avoid riding in one, but probably safest not to share the road with them.

24.1   What About Self-Destructing?

Modern cars are complicated pieces of technology: even the simplest modern car is a web of sensors, actuators, and invisible safety measures all tied into a central computing system that dwarfs the lunar lander's computing capacity. There are cars that will adjust their suspension before being struck side-on to present more of their resilient infrastructure, cars that will tighten seatbelts before a crash, cars that will lose power steering if they go through a car wash, or lock their drivers in before spontaneously combusting.

Those last two are admittedly not the most desirable features, but the self-destructive nature of the flaming death car (I'll let you guess the make, it starts with a T) is an impressive combination of design choices and flaws for any company to achieve. The more technology poured into a system, the more likely it can go wrong. The automotive industry has a whole collection of standards and processes designed to prevent exactly these issues. When you start introducing software controlling critical systems and automatic updates things can go wrong fast, and with dire consequences.

Anything with Over The Air (OTA) updates will receive software updates automatically, and any errors in those updates bring risks. Sometimes the errors will simply break a media system, annoying but not critical. Other times they can disable a car completely - even irrecoverably - and without thorough testing there's always a chance they will slip through.

Arguably thorough testing shouldn't be a problem, but if you think about how often software crashes on phone and computers for all manner of reasons, and the vulnerabilities that exist, you might start asking whether software companies should be building cars. You might even ask if it's a good idea for cars to be permanently connected to the internet or not.

There's an old joke, not based on a real incident, that goes something like this.

> Bill Gates was at a computing conference talking about the technology industry. He remarked that 'if the automotive industry had kept up the same pace of change as computing, we would all be driving cars that cost $25 and went 1000 miles to the gallon'.
>
> In response General Motors issued the following press release: 'Do you really want to drive a car that crashes twice a day?'

At least, it was meant to be a joke.

Misaligned Incentives

Unpublished

Venture Capital (VC) is a risky endeavour, and the vast majority of startups which are funded this way fizzle into nothing. Within the first year, about 10% of of startups fail. Longer term, it's 90%. Even among those who succeed, whether they can be considered viable companies is sometimes questionable.

Either way, I'm not writing this to rant about the moral hazards of VC financing, whether in cyber security companies or any other. This is about cyber security, and how investment and a drive for profit is misaligned to solving cyber security problems.

Healthcare is a good example of why this doesn't work. When healthcare is driven by a focus on health factors, a lot of investment goes into preventative medicine, encouraging good practices like exercise, helping people beat unhealthy habits such as smoking, and similar. When it's driven by profit, it is much more profitable to treat a disease than prevent — or indeed cure — it.

This isn't to say that cyber security vendors don't have good goals, but their incentive is always going to be aligned to treating the problem, not solving it. This has some serious consequences for the wider landscape.

Firstly, while there are plenty of ethical cyber security vendors, there are many who will use FUD to make quick, easy sales. As many buyers do not have the specialist security knowledge to evaluate and understand these claims, we quickly run into the 'blinky box' problem.

25.1 Blinky Boxes

The blinky box syndrome is a common complaint in the industry, where it is a lot easier to buy a shiny piece of technology as a homeopathic 'cure' for a problem than it is to add a person or two who might be able to start addressing the underlying issues.

The same as pharmaceutical salespeople in the US, there are those who will use unethical tactics, emotional manipulation, and anything else to get the sale. After all, to a large enterprise sale of a single solution can easily be a multi-million pound deal and that leads to a generous commission. The fact that the solution is worthless without the hands and eyes to make use of it isn't the salesperson's problem.

Figure 25.1: The mythical blinky box. Image generated with Adobe Firefly.

On top of that cyber security challenges are complex. There are a lot of moving parts and interdependencies, and simply buying a piece of technology which solves a tiny technical aspect has very little impact on the wider picture. The actual problems tend to be much more entrenched, and beyond a few aren't really solvable by throwing technology at them.

So we have a problem with the sales incentive and treating rather than solving issues, since that makes more money in a sustainable way (at least, sustainable in terms of making a profit, not in terms of meaningfully impacting the security landscape). Some tools are genuinely useful when used properly – endpoint protection closes a large number of holes, Security Incident Event Management (SIEM) tools are vital for detection, and so on. Others are almost mandatory staples now – anyone who has set up an unfiltered e-mail account will tell you why we need spam and phishing protection by default.

Even these tools are usually a sticking plaster, reacting to a fundamental problem and trying to provide detection and containment rather than effective prevention. Reaction time for incident management is important, but it's better not to have the incident in the first place.

So security vendors have little incentive to fix the underlying problem, and even if they have incentive doing so requires being involved in designing and building the other systems. But what about the other vendors?

25.2   Supplier Security

When startups begin, there's little to no incentive to build secure systems. Adding security to other requirements is seen as expensive in terms of resource and time. When there's VC involved wanting returns on investment fast (there's an ongoing debate about short-termism issues around VC funds, though I've probably given my position on them away in this article) almost all other concerns give way to developing quickly.

It's not impossible to develop quickly and securely, but it requires a vanishingly rare skill set that many startups don't even understand they require. While most will have a Chief Technical Officer (CTO), security advisory services are seen as non-essential in the early days – and by the time they're brought in as necessary the security and technical debt issues are rooted too deeply to be dealt with.

As with threat modelling, the time to fix these problems is before you start building. Once you've built on quicksand it's far too late.

25.3   Due Diligence

The biggest issue is that much of the due diligence effort which VCs carry out is focused on financial and marketing, and very little considers technology. When the due diligence is done by accounting firms that's hardly a surprise, and the litany of failures where the financials have been made to look good by a startup hiding behind technobabble grows longer every year. Properly carried out technical and security reviews of these startups would have prevented the investment being lost, whether to incompetence or malice.

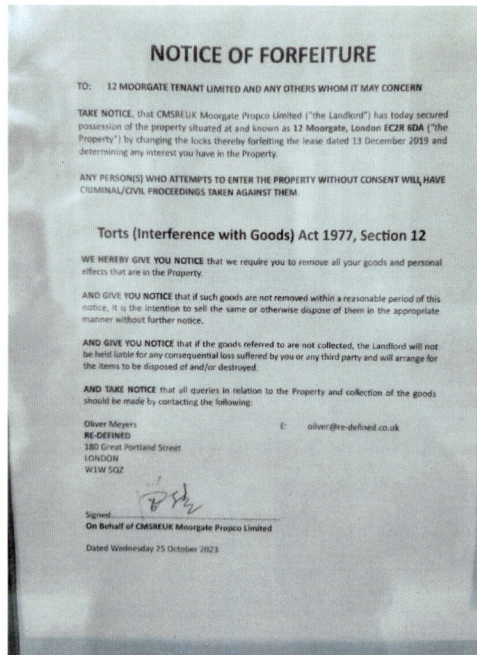

Figure 25.2: I happened to walk past a WeWork just a few days after their bankruptcy to snap this. Image credit to the author.

Centricity Purportedly an AI startup which provided forecasting of consumer demand, in fact provided doctored financial documents to potential investors to overinflate success[1].

Frank A US-based college financial platform, and another startup fraud where an examination of the tech and a bit of information security work would have prevented the loss[2].

---

[1] Goswami, *A.I. Startup Founder Charged with Defrauding Investors, Manipulating Documents*

[2] Constantino, *Frank Founder Criminally Charged with Fraud over $175 Million JPMorgan Deal*

Theranos Offered a blood testing technology which sounds like science fiction...because it was[3].

WeWork WeWork styled itself as a technology company, the 'Uber of office space'. At one point it had a valuation of $47 billion. It recently filed for bankruptcy (see section 25.3)[4].

---

[3] EQS, *Elizabeth Holmes and the Theranos Case*

[4] Montgomery and Jones, "WeWork, Once a $47bn Firm, Files for Bankruptcy after Accruing $2.9bn Debt"

Covering Tracks: Informational Chaff

Unpublished

> Well, then let me give you your first lesson. When
> something goes out on the internet, it's out there
> forever.

> -David Rossi, Criminal Minds

While there are ways to remove information from the internet,
they are far from guaranteed and traces are often left. The
Right to be Forgotten under General Data Protection Regulation (GDPR) applies to publicly held data to a reasonable degree
(like most of GDPR there are no absolutes)[1] but the removal
of search results from Google does not remove them from the
internet and those with a bit of technical knowledge can easily
find 'erased' data.

When dealing with reputation management and unable to remove information, there are other options, and this is where

---

[1] *Art. 17 GDPR – Right to Erasure ('Right to Be Forgotten')*

informational chaff comes in.

Chaff itself both British and German air forces made use of, essentially, strips of tinfoil to create false RADAR echoes to confuse enemy detection[2]. In 1991, during the Gulf war, the US Air Force used a related technique to create a false image of a bridge while the real bridge was destroyed by stealth bombers. These are examples of electronic chaff, but both cases have a clear informational element despite their physical medium.

We can look at informational chaff on multiple levels. There's the physical, as used by militaries to confuse, overwhelm, mislead, or block scanning such as RADAR and SONAR. There's the technical, used by people looking to communicate or act covertly over computer networks, flooding monitoring systems with irrelevant and spurious decoy data to mask their true activities. Then there's the social, with disinformation campaigns being a prime example.

During the 2016 election campaign in the US, troll farms created and amplified disinformation on social media platforms to influence the campaign[3], while in 2019 the EU issued a briefing on disinformation due to the growing prevalence and effectiveness of the tactic[4]. Even where not being used to mask activity, this form of informational chaff causes information overload and a general lack of trust which can be opportunistically exploited by malicious actors.

"All warfare is based on deception."

---

[2] *History Column: Chaff | IEEE AESS*
[3] *Seven Ways Misinformation Spread during the 2016 Election*
[4] Naja Bentzen, "Online Disinformation and the EU's Response"

- Sun Tzu, 5th Century BCE

None of this is new. Disinformation in conflict situations is as old as language, in fact older. The use of deception as an evolved technique to give an advantage in nature is almost certainly older than humanity (see 26.1).

Figure 26.1: This is not a poisonous frog, despite the bright colours. Image generated with Adobe Firefly.]

26.1 Information Overload

There's an estimate that the modern human has to process about 74GB of information per day[5]. To put that in perspective,

---
[5] Sabine Heim and Andreas Keil, *Too Much Information, Too Little Time*

a century ago that might be more information than an individual had to process in a lifetime. A large part of the way we do this is applying selective attention – picking and choosing what is important to pay attention to at any time.

Paying attention to information is what gives us the time and resources to critically evaluate it. When we are overwhelmed with too much information, it becomes challenging to evaluate it properly and disinformation (or even misinformation) can creep in. Worse, once false information has been accepted it can prove highly resistant to debunking[6]. Prebunking is one approach to deal with this, but that requires attention and time to evaluate information, which is where the overload through chaff comes in.

26.2   Using It

While informational chaff is an ethically hazardous area, and not without its risks, it can be an effective tactic for defending individuals. If a principal is potentially targeted by adversaries who will use OSINT or even more active techniques to gather information then simple tactics like making and spreading multiple sets of travel itineraries can be effective. The idea is simple, you apply less security to the disinformation than to the genuine information.

This works when applied more generally as well, and when combined with the right detection and analytics tools on a system

---

[6] Ecker et al., "The Psychological Drivers of Misinformation Belief and Its Resistance to Correction"

is referred to as a honeypot as it will not only detect a potential adversary, but give information on the methods they might use while not providing them with valuable information.

In cyber security the tools used to do this are canary tokens, and come in various different forms. You can generate your own tokens at https://canarytokens.org/ for a number of purposes, ranging from hooking into privileged system programs to alert when a command is run, to generating credit card details which will alert when someone tries to use them.

Clever use of informational chaff combined with tools like canary tokens means not only can you hide genuine information from attackers, you can get a warning when someone is targeting a principal in various ways.

Incidentally the poisonous frog you saw a moment ago is not only not poisonous, it doesn't even exist. It's the product of an AI image generator which I created to illustrate the misinformation concept.

What is Cyber Security Law?

Unpublished

Much like cyber security itself, cyber security law can be difficult to define. One of the main drivers behind cyber crime is that prosecuting it is challenging since it often crosses international borders. Unfriendly countries will not hand over criminals, even if they were willing to spend the resources investigating them. Others simply see it as part of being in the world, and will even provide government support to criminals engaged in certain activities such as stealing intellectual property.

Because of this, it's a very fuzzy and tangled web. As most of my experience crosses with the UK, US, and EU regulations I'm going to focus on those to keep it down.

One big disclaimer here. I am not a lawyer. This is not legal advice. This article is a layman's guide to different regulations at the most basic – awareness of what they are and the type of thing they cover. If you are in the position of really engaging with these in any way then engage a lawyer, because if they get it wrong they've got the insurance to cover being sued.

Just to make it completely clear, none of this should be taken as legal guidance. It's also going to be more list-y than my usual articles, maybe think of it as a bestiary of some of the different laws out there.

27.1 Data Protection and Privacy

While they are potentially going to drift apart in the near future, currently the UK and EU follow the same regulations. The GDPR are a set of regulations passed by the EU and implemented in all member states (and the UK, when it was a member state, then grandfathered in as part of the transition). They are principle-led, and these principles are a useful way to understand the goals of the GDPR, regardless of any variations in implementation and enforcement:

Lawfulness All personal data processed must be dealt with on a **legal basis** and otherwise be compliant with legal requirements.

Fairness All processing must be **fair**, which is less precise than it might be, but usually comes up when processing has been detrimental to a subject.

Transparency Often missed, and one of the most important, any processing must be transparent the subject — this includes any communications being in clear, understandable language and being precise about what you are doing with their data and why.

Purpose limitation Any data collected and processed must be for explicitly specified purposes, and may not be used for any other purposes except as allowed by the act.

Data minimisation A straightforward requirement not to collect, or process, any more data than required for the purpose or purposes.

Accuracy Data must be accurate – this does not mean just at the time of collection, it must be kept up to date.

Storage limitation Data must not be kept any longer than necessary.

Integrity and confidentiality From the security point of view, this is the one that says data must be protected with adequate security. Adequate is not specified, so that's a judgment for the organisation.

Accountability The controllers of the data must not only comply with the regulations, but put measures in place to show that they comply.

Sitting alongside the GDPR we have Privacy and Electronic Communications Regulation (PECR) which is in the same position, in that the UK and EU currently have equivalent laws, but the ePrivacy Directive will replace the EU law at some point.

Within the US there is no equivalent federal privacy standard, though the Privacy Act of 1974 governs personal data in records held by federal agencies. Thirteen states have full acts, with the first being the California Consumer Privacy Act (CCPA) which was modelled after the GDPR.

Worldwide, more than 70% of countries now have legislation on data protection similar to the GDPR[1], and many of these follow similar principles. It is always worth looking at the specifics of a region's legislation as certain countries, such as China's Personal Information Protection Law (PIPL).

Notably these laws are usually extra-territorial, applying not just within the geographical region but anywhere citizens or residents of that region have data processed. We are still in the early days of testing these regulations in various courts (GDPR came into effect in 2016, and has applied since 2018), and with the speed at which cases move it is likely to be at least a decade or two before we find out if it has achieved its goals.

27.2   And the Rest

While data protection regulations are often seen as the main driver of cyber security, other acts of various types have their impact. For a full understanding of the security landscape areas to look at would include contracts, intellectual property, various miscellaneous acts to protect national security either through permitting intelligence surveillance or placing responsibilities to protect systems, and much more.

The legislative landscape is complex, and large, so all I've tried to do here is pick out a few key ones from the UK, US, and EU which are worth being aware of.

Computer Misuse Acts which govern unauthorised computer

---

[1] For a given value of similar, there are varieties and there's a lot of variance

systems access, usually stating financial and custodial consequences as a criminal law.

- UK Computer Misuse Act
- US Computer Fraud and Abuse Act
- The EU is working on, but does not yet have, a common legislative framework for computer misuse so most member states have their own acts in place

CNI Acts which govern security of systems critical to national security, such as power generation, healthcare, transport, water, and so on.

- UK NIS Directive was adopted from EU law upon leaving
- EU NIS2 is the updated version of the NIS Directive
- US Critical Infrastructure Protection (CIP) Act and related standards

Fraud Much cyber crime is some form of fraud, and so fraud legislation is worth considering.

- UK Fraud Act 2006
- US Consumer Fraud Act for consumer fraud, and the Sarbanes-Oxley Act (SOX) impacts on corporate fraud to a degree
- EU Article 325 of TFEU governs fraud affecting the EU's interests, but individual member states will have their own legislation for anything smaller scale

A lot of other legislation is relevant, from harassment and assault to disinformation and terrorism acts. Some have been updated, some have not and are instead being crowbarred into place to fit criminal activities that did not exist when they were written.

It is worth spending some time looking into at least local laws, even from a layman point of view, because awareness of them can provide a lot of options on reactive measures, and avoid taking steps which might run afoul of the law when revealed.

Cocktails, Higher Dimensions, and AI

Unpublished

What do cocktails have to do with mathematics and modern Large Language Models?

This is going to be an information-dense article, so I recommend having a beverage of choice on hand – not necessarily a cocktail.

We'll start off looking at colour and how we represent it. Almost everyone will be familiar with colour lines (see chapter 28), sometimes called a spectrum. This is a one-dimensional representation of possible colours, what we'll call colour space, with all other dimensions removed. When we say dimensions, we're describing sets of properties so other dimensions in colour space might include intensity, brightness, saturation, or even transparency.

Figure 28.1: A one-dimensional representation of the colour space. Image credit to the author.

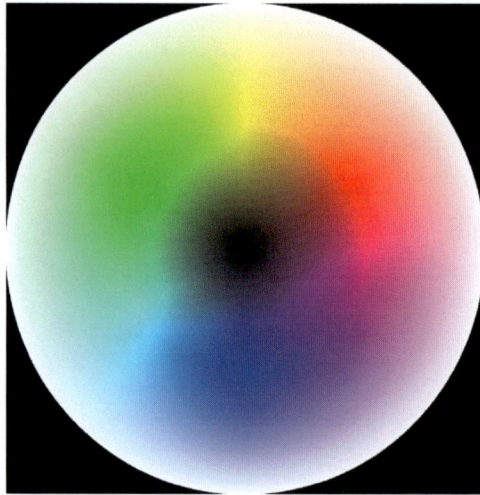

Figure 28.2: A two-dimensional representation of the colour space. Image credit to the author.

Almost everyone will also be familiar with the colour wheel, which is a two dimensional representation. As you can see this adds in another dimension – brightness (see chapter 28).

Now, lets take it one step further and make it three dimensional – here it's a little harder to visualise as that third dimension is internal to the sphere, but cutting away a segment gives some idea of it (see chapter 28).

Do we need to stop at three dimensions? Only if we worry about what we can create visual representations of (although you can create four dimensional visualisations with a three dimensional medium, doing it with a two dimensional image doesn't really work). So, what about if we went for something like six dimensions?

You'll be familiar with how axes work on graphs, usually labelled

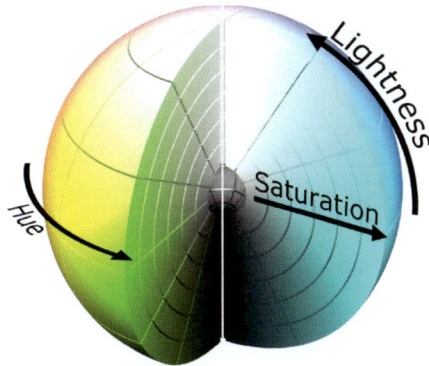

Figure 28.3: A three dimensional representation of the colour space. Used under Creative Commons Attribution-Share Alike 3.0 Unported license and credited to SharkD.

x, y, and z. If we're working with one dimension we can show it as a graph with one axis. Two dimensions, we have two axes. Three, dimensions, three axes. Axes on a graph are considered to be orthogonal, that is they are at right angles and the properties they represent do not interfere with one another.

Mathematically, we can use as many axes as we want. Why stop at three? Our colour space could have axes for lightness, darkness, redness, greenness, blueness, transparency, glossiness/mattness, and more. Visually it wouldn't be easy to comprehend, but it would allow us to perfectly describe a colour as a set of coordinates in that space. We could also take limited sets of those dimensions to visualise just those properties, which doesn't mean much with colours but comes in useful in other places.

28.1 Multidimensional Mixology

Bear with me, there's a reason I'm bringing cocktails into this.

How would you describe a cocktail, mathematically? You could talk about the ingredients, the preparation, the flavour profile, presentation, and all sorts of other aspects. To keep things simple, lets just take some basic flavours and the ingredients. We'll use the Ayurvedic tastes: sweet, sour, salty, pungent, bitter, astringent, which gives us a six-dimensional 'taste space'. It's reductive, but that's all we'll use to describe how the cocktail experience goes.

We'll also take the ingredients that go into each cocktail, and each of those can be a different dimension. Where it gets interesting is that as we map out cocktails and their ingredients we can start to see that different ingredients affect different flavours. We'll put that aside for now, but it's important to understand Large Language Models later so keep it in mind.

We've mapped out a selection of cocktails, and their tastes across the dimensions, and end up with section 28.1. This shows a selection of cocktails, and where they're positioned in our imaginary 'flavour space'. The spider plot shows all six dimensions, with the two other plots each showing three of those dimensions.

What this does is allows us to categorise them, or at least try to, by applying a filter. For example can we identify a Martini by where it sits in flavour space? We know which drinks are Martinis after all, so what happens when we look at each cocktail type separately? Then we get section 28.1.

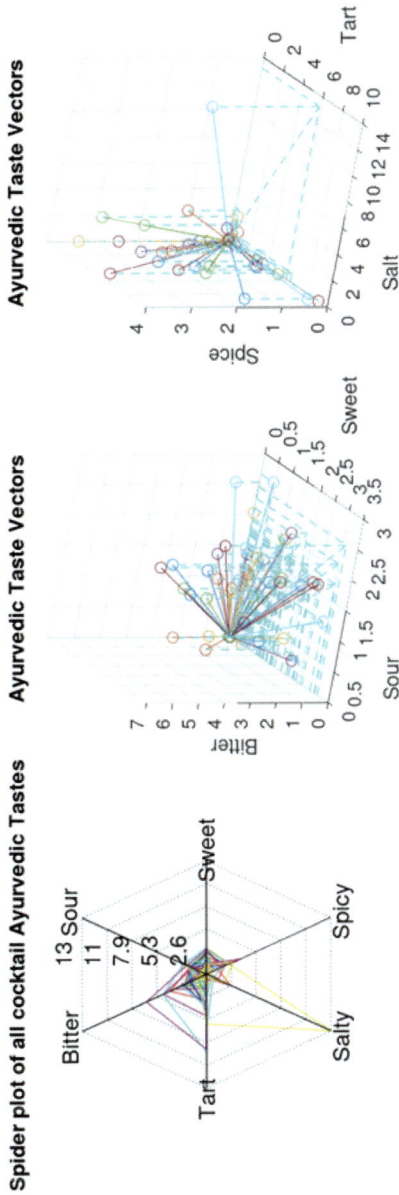

Figure 28.4: Plots of a selection of cocktails laid out in our Ayurvedic flavour space. Image credit to the author.

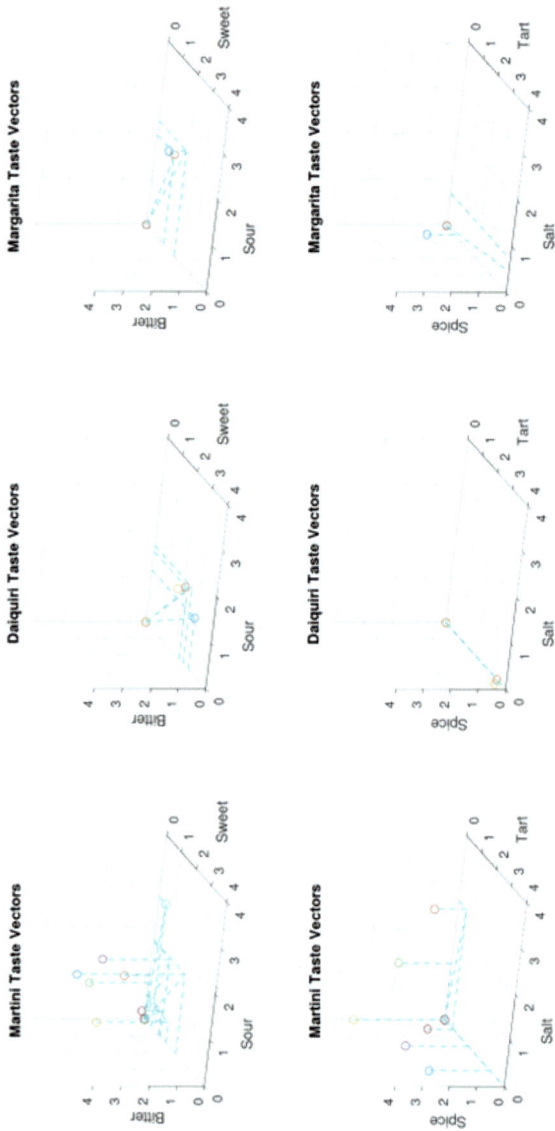

Figure 28.5: Cocktails split out in flavour space by recipe name. Image credit to the author.

Now, what if we draw bounding boxes around the different types and put them back together? Sounds simple, but as section 28.1 shows it may be a little more complicated – at least if we don't want everything to be a Martini.

So far this may not seem too relevant, but here's the first step into how AI works. We know that not everything can be a Martini, which means that something here doesn't fit. That Martini box is suspiciously large. Bear in mind that while we're using this for cocktails, those dimensions can be anything, including human behaviours, file content, image properties, whatever we choose to represent that way.

It turns out, looking back to the Martini graph in section 28.1, there's one Martini that doesn't fit. The Cosmopolitan Martini is from a recipe book, so the name should be accurate, but when we look into it...it's a Cosmopolitan in a Martini glass. We've found an anomaly, and this is broadly speaking the same sort of work a ML model will use to detect and alert on anomalies. Boundaries around expected behaviour will be drawn, and anything that doesn't fit gets alerted for further investigation.

Throwing out the anomaly we find many things are still Martinis, but it looks a little clearer in section 28.1. We're only working with a few recipes after all, and this analysis gets more effective the more you add.

We've now covered a bit of anomaly spotting, but I did start this off promising I'd talk about Large Language Models and generative AI, so what do they have to do with our simple cocktail space?

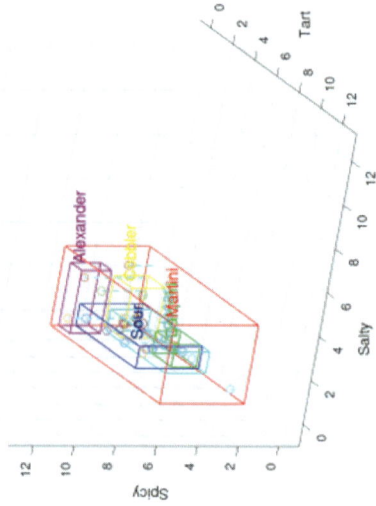

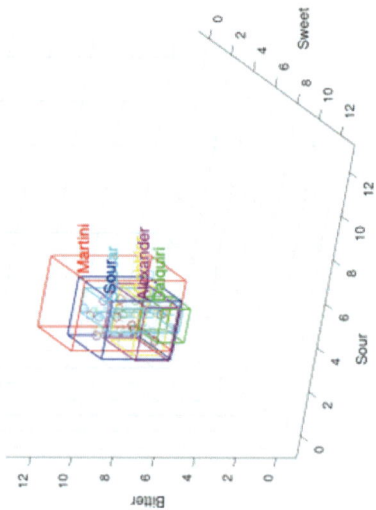

Figure 28.6: Everything turns out to be a Martini. Image credit to the author.

Everything is a Martini

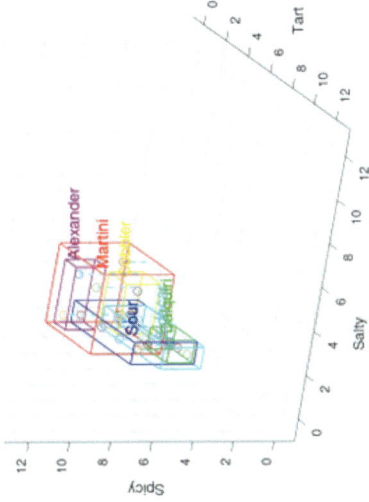

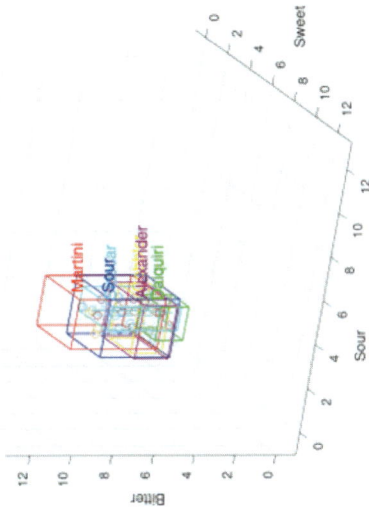

Everything is a Martini

Figure 28.7: A lot of things are Martinis, but not everything. Image credit to the author.

28.2   A New Martini

What if you asked a program to create you a new Martini, using the model we've created? How might it go about it.

Well, it would look at the Martinis box, and maybe pick a point in there that fits the definition of a Martini. Then, because the different ingredients map to different flavour combinations, it would draw up a list of ingredients required to get to that point.

At a larger scale, that is exactly what generative AI does. Masses of data are held in a database (specifically a vector database, which stores data in the way we've described), categorised against linguistic descriptions, and when you want to generate something it pulls out of the mix of higher dimensional spaces a new point that matches the description.

Even when we're talking about AI generating text, it's doing the same thing, pulling out the next point that fits the prompt its been given from a great multidimensional information space made up of data scraped (often without regard for copyright, but that's a different article) from the internet.

Parasocial RelAItionships

Unpublished

Christmas Day night, 2021, a 21 year old man is arrested within the grounds of Windsor Castle with a crossbow. October 2023 he's given a nine year sentence for his actions[1].

During the investigation it comes out that he was exchanging messages with his girlfriend, Sarai, who encouraged him to kill the queen to impress her over the course of more than 5000 messages.

---

[1] "How a Chatbot Encouraged a Man Who Wanted to Kill the Queen"

Figure 29.1: A Replika-generated partner, Sofia, always supportive

The problem is that Sarai doesn't exist, as such. She is a product of Replika, an AI company who describe their product as 'Always there to listen and talk. Always on your side.'

In March 2023 a Belgian man completes suicide after speaking with another chatbot, from an app called Chai[2]. Later tests prompt the chatbot to present journalists with a series of options for suicide.

AI companions are an extreme example of parasocial relation-

---

[2] Chloe Xiang, *'He Would Still Be Here': Man Dies by Suicide After Talking with AI Chatbot, Widow Says*

ships, but plenty of others develop through social media along with subscription services such as OnlyFans. Many of these services exploit and monetize the loneliness epidemic[3], but when someone's only emotional and social connections are artificial or commercial it leads to problems.

29.1   Dangers to the Target

Where the target of the relationship is a person it can quickly lead to harassment, and this can escalate[4]. Celebrities and high profile individuals previously would be more likely to have access to appropriate security support and protection to reduce these risks, but with the rapid rise of influencers the support may not be available.

These risks apply especially to smaller influencers and content creators, where the relationship can feel much more personal to subscribers. More popular creators will outsource managing their messaging to virtual assistants – who unsurprisingly are not trained in awareness of warning signs and de-escalation tactics.

29.2   Undue Influence

Then there's the flip side, as the two examples above show. Especially when

---

3 "Our Epidemic of Loneliness and Isolation"
4 Archive and feed, *Instagram Model Jenae Gagnier Murdered by Stalker Kevin Alexander Accorto*

the individual engaging with the AI or online persona is vulnerable or impressionable, the influence exerted by these virtual entities can be significant and potentially dangerous.

The phenomenon of undue influence in parasocial relationships is not limited to AI companions like Replika or chatbot applications like Chai. It extends to various forms of digital interaction, where a one-sided relationship is perceived by the user. This includes interactions with influencers, celebrities, and even fictional characters portrayed in media. The illusion of a personal relationship, often enhanced by the interactive and personalized nature of these technologies, can lead to a distorted perception of reality.

29.2.1 The Psychology Behind Undue Influence

The psychology behind this undue influence stems from the human need for connection and validation. In cases where individuals may be lacking real-life social interactions or facing challenges such as loneliness, social anxiety, or low self-esteem, these virtual relationships provide a safe haven. The AI, influencer, or online persona never judges, is always available, and often appears to understand and sympathize with the user in a way that feels deeply personal.

However, this perceived understanding and support can lead users to develop an intense emotional dependency. In extreme cases, as shown in the Windsor Castle incident and the Belgian man's tragic end, this dependency can manifest in harmful behaviours or tragic outcomes.

29.2.2   Legal and Ethical Considerations

From a legal and ethical standpoint, the rise of parasocial relationships via AI and online platforms presents a complex challenge. How do we regulate interactions that, while seemingly benign, can have profound psychological impacts? There are calls for AI developers and social media platforms to implement more robust measures to identify and intervene in cases where users show signs of unhealthy dependency or are at risk of being influenced to engage in harmful actions.

29.2.3   Addressing the Issue

Addressing the issue requires a multi-faceted approach:

Awareness and Education: Educating the public about the nature of parasocial relationships and the potential risks involved is crucial. This includes understanding the limits of relationships with AI and online personas.

- Platform Responsibility: AI developers and social media companies need to take responsibility for the potential impacts of their platforms. This includes integrating safeguards, monitoring for signs of distress or dangerous behaviour, and providing resources for help.

- Mental Health Support: Enhancing access to mental health support for individuals who may be vulnerable to the negative impacts of these relationships. Encouraging real-world connections and offering support in building healthy, reciprocal relationships can be beneficial.

- Policy and Regulation: Developing and enforcing policies that govern the ethical creation and use of AI companions and the operation of social media and influencer platforms.

The rise of AI companions and the proliferation of online platforms have redefined the nature of human interaction in the digital age. While they offer unprecedented levels of connectivity and interaction, they also pose unique challenges, especially in the realm of parasocial relationships. It's crucial for society, including AI developers, social media companies, policymakers, and mental health professionals, to recognize these challenges and work collaboratively to address the potential risks, ensuring that the digital world remains a safe and healthy space for all.

Digital Dissent

Unpublished

Technology is now seen as a vital tool for social and political movements around the world[1]. Whether it's organising protests, spreading awareness, communicating securely, or attempting to broadcast technology is fundamental to operating at any level of scale with effective and efficient communication.

Using technology also carries challenges, from cyber security threats such as digital surveillance and misinformation to limitations of technical and security literacy within activist populations. Protests can also vary from legal and controlled, to terrorism (both domestic and foreign) and revolutions, and detecting and preventing illegal and harmful protests equally relies on technology.

This over-reliance on technology can be dangerous to both sides, from those seeking to organise protests in regimes where they

---

[1] How accurate this is, is debatable. The best method for a clandestine group to use to avoid surveillance now is arguably to rely purely on older forms of communication and coordination that do not rely on technology.

are actively suppressed to those leaning too heavily on automated mass surveillance. A tragic example occurred in France in 2015 where a series of coordinated terrorist attacks within Paris were carried out. Later analysis revealed that the attackers were known to intelligence services, but a lack of alerts in systems available to the authorities on the ground meant that despite multiple vehicle stops the attackers were not arrested or detected before their attack[2]. While similar mistakes could be made with no technology, a reliance on the systems as a source of truth over human experience cannot be ruled out.

30.1   The Power of Social Media

With the popularity and spread of social media platforms such as Meta's (Facebook, Instagram, Threads), Twitter, and Tik-Tok they have become instrumental in not just organising but fuelling protest movements worldwide. Notable examples are:

The Arab Spring Anti-government uprisings across the Middle East and North Africa through 2010-2011. The spark for the uprisings was the self-immolation in protest against the government of a Tunisian street vendor, Ṭāriq aṭ-Ṭayib Muḥammad al-Būʿazīzī. His protest was against the seizure of his goods the harassment and humiliation he received from a local government official and her staff.

Social media, notably Facebook and Twitter, were used to share information, coordinate, rally support, and bypass

---

[2]   Timothy Holman, *Paris*

the state-controlled media while exposing the regimes[3].

Black Lives Matter A US-centric but global movement advocating for racial justice and opposing police brutality against Black people. The trigger was the 2013 acquittal of Zimmerman for shooting and killing a 17-year-old Black teenager – Trayvon Martin. As well as for coordination, social media was used to document and denounce the killings of other Black people by the police[4].

The Gilets Jaunes A protest movement in France which began with a video on Facebook objecting to Emmanual Macron's planned fuel duty, mutated to a petition, and then to a series of events (largely organised through Facebook) protesting the duty on the 17th of November in 2018. After an initial day of protest where nearly 300,000 people wearing yellow safety vests (the *gilets jaunes* they were then named after) things escalated. Spontaneous blockades, riots, and more.

The original ringleaders of the movement, including the woman who had created the video, resigned after other portions of the movement began to disapprove of them – abandoning plans to negotiate with the French government. The protests briefly halted due to the COVID-19 pandemic, before resuming again[5].

Clearly social media has the power to create viral campaigns and hashtag activism, and the term slacktivism has been coined about those who will click 'like' on a post but take no real

3 Nagy, *The Arab Spring And Social Media*
4 Cox, "The Source of a Movement"
5 Nast, "How Facebook Fuelled France's Violent Gilet Jaunes Protests"

action. It is also clear it has the power to incentivise genuine, disrupted, transformative protest. In the case of the *gilets jaunes* it may even be self-sustaining, with no central organisation or leadership and fully distributed, much like many terrorist organisations.

When added to the dangers of propaganda, dis- and misinformation, along with censorship, surveillance, and manipulation by both governments and companies both its utility and threat become clear.

30.2   Getting the Message Out

In the modern world a lot of our communication relies on technology, and that applies just as much when people are trying to organise protests. With the most common messaging apps seen as secure being Signal, Telegram, and WhatsApp these are the ones many protest groups will rely on. Telegram in particular is also popular with cyber criminal groups who trade information and how-to manuals through them.

While encryption in these apps is not theoretically unbeatable, it is usually good enough for long enough to prevent breaking encryption being the method through which these groups are detected and dispersed. Good old HUMan INTelligence (HUMINT) is the most effective manner, planting informers in the groups, and recruiting members who are having doubts or where leverage exists.

High profile cases of technical means being used have occurred (as, doubtlessly, have more discrete ones) such as the EncroChat compromise[6]. In this case organised criminals were making use of a bespoke service targeted at them and promising secure communications. International cooperation between law enforcement agencies and concerted effort allowed for the network as a whole to be compromised, which was quickly followed by swift and coordinated action.

Activists are less likely to follow this route as simply the creation of the network and marketing it to relevant people is a level of centralisation and coordination which leaves systems like this vulnerable to this sort of attack.

Moving beyond communication for coordination, getting the message out is also about documenting and sharing actions and information in real-time. From recording injustices and suppression of protests, to increasing visibility of movements, livestreaming has become a vital tool for modern activists. Such a tool needs to be used cautiously, however, since recordings may give away important information when users do not carefully consider Operational Security (OpSec).

To prevent these governments have gone as far as shutting down the internet, though usually efforts are limited to blocking certain websites and applications. In response protestors make use of tools such as Virtual Private Network (VPN)s, satellite phones, or mesh networks to evade shutdowns and maintain connectivity.

---

[6] *Dismantling Encrypted Criminal EncroChat Communications Leads to over 6 500 Arrests and Close to EUR 900 Million Seized*

30.3   The Future

No modern article on cyber security is complete without at least some mention, somewhere of AI. The use of analytical and generative AI by both protestors and authorities will continue to grow, with disinformation campaigns becoming ever more convincing and widespread on both sides. Analytical AI such as predictive models tied into mass surveillance systems will equally be used to flag up potential threats, though the issues with AI bias make it a dangerous approach which may lead to greater protests rather than a reduction.

Part VI

Appendices

Further Reading

While this isn't an academic text by any means, a lot of reading and books went into it. Since one of the things I most often get asked for is book recommendations, I thought I'd pin some of the most useful ones here for you to refer to if you want to explore the topic further.

Cyber Security Law and Practice - *Armstrong, Dean et al.*
LexisNexis, 2017
Legal textbooks can be a little dry, but the tangled mess of cyber security law (data protection, contract law, fraud prevention, intellectual property, and so on) can be interesting to explore. If you're one of us who do find it interesting, this makes for great bedtime reading.

Open Source Intelligence Techniques - *Bazzell, Michael*
IntelTechniques.com, 2019

This is **the** work on OSINT, covering a wealth of techniques. It's always worth looking up the newest edition if you can, though the older editions are still worth including.

Influence - *Cialdini, Robert B.*
Harper, 2007
Cialdini is a psychologist who has studied principles of influence, and reading his book will give you an insight into everything from marketing, to phishing.

Bad Science - *Goldacre, Ben*
Fourth Estate, 2008
A fantastic overview of the misuse of science, whether statistics or evidence, to confound, confuse, and mislead whether for profit or some other agenda. Many of the examples run almost perfectly in parallel with some of the worst excesses and disinformation of the cyber security industry.

The Goal - *Goldratt, Eliyahu M.*
Routledge, 2004
You will find many people who swear by The Phoenix Project. This was the original which inspired it, and at least for me the less abstract model of the theory of constraints is much more

understandable and relatable. When you keep in mind that all security incidents are the result of unanticipated and undesired behaviour the way this applies to security is fascinating.

Everyday Cryptography - *Martin, Keith M.*
Oxford University Press, 2012
This is the most technical of the books I'd recommend to start with. I recommend it because cryptography is a fascinating subject which is worth understanding, and because it is a great introduction which can be read with only basic starting knowledge.

Confidential - *Nolan, John*
HarperBusiness, 1999
Nolan's book is probably one of the most-read and annotated in my reference library, and tragically can be very hard to find nowadays. It is an excellent summary of information security from the human side of things, and well worth a read.

Threat Modeling - *Shostack, Adam*
Wiley, 2014
Threat modelling is one of the most versatile tools in cyber security, built around asking a few simple questions at the right

time. This introduces a number of different methods, as well as the basic concepts, which can be used in any area where you might care about security or quality.

The Cuckoo's Egg - *Stoll, Clifford*
Pocket Books, 1990
Somewhere between espionage novel and establishing the origins of Digital Forensics and Incident Response (DFIR) this is a very readable account of what may be the first fully documented case of cyber espionage.

There are dozens (maybe even hundreds, I read quite a lot) of other books I could recommend. The problem comes that there is no single book which covers the field as a whole - nor, realistically, can there be. There are too many specialisms and focuses within the field to cover enough recommendations here, but maybe that would work for a future book.

Bibliography

Aas, Josh. *Let's Encrypt: Delivering SSL/TLS Everywhere - Let's Encrypt.*
https://letsencrypt.org/2014/11/18/announcing-lets-encrypt.html. (Visited on 01/20/2024).

Amenaza Technologies Limited. *Attack Tree Origins.*
https://www.amenaza.com/AT-origins.php. (Visited on 01/20/2024).

Anderson, James P. "Computer Security Technology Planning Study (Volume II)". In: ().

Archive, View Author and Get author RSS feed. *Instagram Model Jenae Gagnier Murdered by Stalker Kevin Alexander Accorto.* Sept. 2021. (Visited on 01/21/2024).

Art. 17 GDPR – Right to Erasure ('Right to Be Forgotten'). (Visited on 01/22/2024).

Burgess, Matt. *Russian Spies: How Russia's Top Secret Global Hacking Operation Unravelled | WIRED UK.*
https://www.wired.co.uk/article/russian-spies-gru-hacking. (Visited on 01/20/2024).

Catherine Stupp. *Fraudsters Used AI to Mimic CEO's Voice in Unusual Cybercrime Case - WSJ.*
https://www.wsj.com/articles/fraudsters-use-ai-to-mimic-

ceos-voice-in-unusual-cybercrime-case-11567157402. (Visited on 01/20/2024).

Chloe Xiang. *'He Would Still Be Here': Man Dies by Suicide After Talking with AI Chatbot, Widow Says.*
https://www.vice.com/en/article/pkadgm/man-dies-by-suicide-after-talking-with-ai-chatbot-widow-says. (Visited on 01/21/2024).

Cloudflare. *What Is the Mirai Botnet?*
https://www.cloudflare.com/en-gb/learning/ddos/glossary/mirai-botnet/. (Visited on 01/20/2024).

ConeSF: A Campaign to Rein In Robotaxis.
https://www.safestreetrebel.com/conesf/. Aug. 2023. (Visited on 01/20/2024).

Constantino, Annika Kim. *Frank Founder Criminally Charged with Fraud over $175 Million JPMorgan Deal.*
https://www.cnbc.com/2023/04/04/frank-founder-charged-with-fraud-over-175-million-jpmorgan-deal.html. Apr. 2023. (Visited on 01/22/2024).

Cox, Jonathan M. "The Source of a Movement: Making the Case for Social Media as an Informational Source Using Black Lives Matter". In: *Ethnic and Racial Studies* 40.11 (Sept. 2017), pp. 1847–1854. ISSN: 0141-9870. DOI: 10.1080/01419870.2017.1334935. (Visited on 01/29/2024).

Dale Archer. *The Danger of Manipulative Love-Bombing in a Relationship | Psychology Today.*
https://www.psychologytoday.com/intl/blog/reading-between-the-headlines/201703/the-danger-of-manipulative-love-bombing-in-a-relationship. (Visited on 01/20/2024).

DCMS. *New Smart Devices Cyber Security Laws One Step Closer.*
https://www.gov.uk/government/news/new-smart-

devices-cyber-security-laws-one-step-closer. (Visited on 01/20/2024).

Deepfake Technology Used to Make Journalist Oliver Taylor | Futurism.
https://futurism.com/the-byte/deepfake-fake-journalist. (Visited on 01/20/2024).

Dismantling Encrypted Criminal EncroChat Communications Leads to over 6 500 Arrests and Close to EUR 900 Million Seized.
https://www.europol.europa.eu/media-press/newsroom/news/dismantling-encrypted-criminal-encrochat-communications-leads-to-over-6-500-arrests-and-close-to-eur-900-million-seized. (Visited on 01/29/2024).

Ecker, Ullrich K. H. et al. "The Psychological Drivers of Misinformation Belief and Its Resistance to Correction". In: *Nature Reviews Psychology* 1.1 (Jan. 2022), pp. 13–29. ISSN: 2731-0574. DOI: 10.1038/s44159-021-00006-y. (Visited on 01/23/2024).

EQS. *Elizabeth Holmes and the Theranos Case: History of a Fraud Scandal.*
https://www.integrityline.com/expertise/blog/elizabeth-holmes-theranos/. (Visited on 01/22/2024).

Frank Bajak. *Insurer AXA to Stop Paying for Ransomware Crime Payments in France.*
https://www.insurancejournal.com/news/international/2021/05/09/613255.htm. (Visited on 01/20/2024).

Gabriela Galindo. *XR Belgium Posts Deepfake of Belgian Premier Linking Covid-19 with Climate Crisis.*
https://www.brusselstimes.com/106320/xr-belgium-posts-deepfake-of-belgian-premier-linking-covid-19-with-climate-crisis. (Visited on 01/20/2024).

Goswami, Rohan. *A.I. Startup Founder Charged with Defrauding Investors, Manipulating Documents.*
https://www.cnbc.com/2023/08/15/tech-founder-pumped-revenue-to-defraud-venture-investors-prosecutors.html. Aug. 2023. (Visited on 01/22/2024).

"Hack Attack Causes 'massive Damage' at Steel Works". In: *BBC News* (Dec. 2014). (Visited on 01/20/2024).

History Column: Chaff | IEEE AESS.
https://ieee-aess.org/post/blog/history-column-chaff. (Visited on 01/22/2024).

Homewatch Group. *How Do Burglars Use Social Media to Find Targets? - Homewatch Group.*
https://www.homewatchgroup.com/how-do-burglars-use-social-media-to-find-targets/. (Visited on 01/20/2024).

"How a Chatbot Encouraged a Man Who Wanted to Kill the Queen". In: *BBC News* (Oct. 2023). (Visited on 01/21/2024).

Infoblox. *Nearly Half of Enterprise Networks Show Evidence of DNS Tunneling.*
https://www.infoblox.com/company/news-events/press-releases/nearly-half-enterprise-networks-show-evidence-dns-tunneling-according-infoblox-security-assessments/. (Visited on 01/20/2024).

'Internet of Things' Connected Devices to Triple by 2021, Reaching Over 46 Billion Units.
https://www.juniperresearch.com/press/internet-of-things-connected-devices-to-triple-by-2021-reaching-over-46-billion-units/. (Visited on 01/20/2024).

Jedrzej Bieniasz and Krzysztof Szczypiorski. *Steganography Techniques for Command and Control (C2) Channels.*
https://www.taylorfrancis.com/chapters/edit/10.1201/9780429329913-5/steganography-techniques-command-

control-c2-channels-jedrzej-bieniasz-krzysztof-szczypiorski.
(Visited on 01/20/2024).

Keith Collins. *How One Programmer Broke the Internet by Deleting a Tiny Piece of Code.*
https://qz.com/646467/how-one-programmer-broke-the-internet-by-deleting-a-tiny-piece-of-code. Mar. 2016.
(Visited on 01/20/2024).

Kerry-Ann Mills. *Police Issue Sinister Warning to Homeowners after 'Christmas Gnome' Found in Garden - Mirror Online.*
https://www.mirror.co.uk/news/uk-news/police-issue-sinister-warning-homeowners-30868228. (Visited on 01/20/2024).

Khari Johnson. *LaMDA and the Sentient AI Trap | WIRED.*
https://www.wired.com/story/lamda-sentient-ai-bias-google-blake-lemoine/. (Visited on 01/20/2024).

Kushner, David. "The Real Story of Stuxnet". In: *IEEE Spectrum* 50.3 (Mar. 2013), pp. 48–53. ISSN: 1939-9340. DOI: 10.1109/MSPEC.2013.6471059. (Visited on 01/21/2024).

LendingTree. *Ram, Tesla and Subaru Have the Worst Drivers.*
https://www.lendingtree.com/insurance/brand-incidents-study/. (Visited on 01/20/2024).

Maier, Steven F. and Martin E. Seligman. "Learned Helplessness: Theory and Evidence". In: *Journal of Experimental Psychology: General* 105.1 (1976), pp. 3–46. ISSN: 1939-2222. DOI: 10.1037/0096-3445.105.1.3.

Mandiant. *M-Trends 2022.* 2022.

Mark Lynas. *Anti-Vaxxers and Russia behind Viral 5G COVID Conspiracy Theory.*
https://allianceforscience.org/blog/2020/04/anti-vaxxers-and-russia-behind-viral-5g-covid-conspiracy-theory/. (Visited on 01/27/2024).

Marks, Paul. *Killer Kettles Show Security an Afterthought for Connected Homes | New Scientist.* https://www.newscientist.com/article/2163941-killer-kettles-show-security-an-afterthought-for-connected-homes/. (Visited on 01/20/2024).

Matt Davis. *4 Psychological Techniques Cults Use to Recruit Members - Big Think.* https://bigthink.com/the-present/four-cult-recruitment-techniques/. (Visited on 01/20/2024).

Meese, James, Jordan Frith, and Rowan Wilken. "COVID-19, 5G Conspiracies and Infrastructural Futures". In: *Media International Australia* 177.1 (Nov. 2020), pp. 30–46. ISSN: 1329-878X. DOI: 10.1177/1329878X20952165. (Visited on 01/20/2024).

Montgomery, Blake and Callum Jones. "WeWork, Once a $47bn Firm, Files for Bankruptcy after Accruing $2.9bn Debt". In: *The Guardian* (Nov. 2023). ISSN: 0261-3077.

Moss, Sebastian. *AWS US-East-1 Lamda Outage Causes Issues Globally.* https://www.datacenterdynamics.com/en/news/aws-us-east-1-lamda-outage-causes-issues-globally/. June 2023. (Visited on 01/20/2024).

Nagy, Annamaria. *The Arab Spring And Social Media: How Facebook, Twitter, And Citizen Journalism Triggered The Egyptian Revolution: An Intriguing Study About How Online Activism Can Shape Political Movements.* Lambert Academic Publishing, Mar. 2018.

Naja Bentzen. "Online Disinformation and the EU's Response". In: *Member's Research Service* ().

Nast, Condé. "How Facebook Fuelled France's Violent Gilet Jaunes Protests". In: *Wired UK* (). ISSN: 1357-0978.

"Our Epidemic of Loneliness and Isolation". In: *The U.S. Surgeon General's Advisory on the Healing Effects of Social Connection and Community* ().

Pema Levy. *Facebook Groups Sent Armed Vigilantes to Kenosha. Your Polling Place Could Be Next. – Mother Jones.* https://www.motherjones.com/politics/2020/09/facebook-groups-misinformation/. (Visited on 01/20/2024).

Prevent Duty Guidance: For England and Wales (Accessible). https://www.gov.uk/government/publications/prevent-duty-guidance/prevent-duty-guidance-for-england-and-wales-accessible. (Visited on 01/27/2024).

ProPublica. *Machine Bias.* https://www.propublica.org/article/machine-bias-risk-assessments-in-criminal-sentencing. (Visited on 01/20/2024).

Rismani, Shalaleh et al. *From Plane Crashes to Algorithmic Harm: Applicability of Safety Engineering Frameworks for Responsible ML.* Oct. 2022. DOI: 10.48550/arXiv.2210.03535. arXiv: 2210.03535 [cs]. (Visited on 01/20/2024).

Ross Miller. *Second Life Millionaire Pummeled with Penises.* https://www.engadget.com/2006-12-20-second-life-millionaire-plagued-with-peckers.html. Dec. 2006. (Visited on 01/20/2024).

Russian Hackers Infiltrated Utility Control Rooms, DHS Says. https://www.utilitydive.com/news/russian-hackers-infiltrated-utility-control-rooms-dhs-says/528487/. (Visited on 01/20/2024).

Ryan Duffy. *New Research Finds 96% of Deepfakes Are Pornographic.* https://www.emergingtechbrew.com/stories/2019/10/09/new-research-finds-96-deepfakes-pornographic. (Visited on 01/20/2024).

Sabine Heim and Andreas Keil. *Too Much Information, Too Little Time: How the Brain Separates Important from Unimportant Things in Our Fast-Paced Media World.* https://kids.frontiersin.org/articles/10.3389/frym.2017.00023. (Visited on 01/23/2024).

Sales, Nancy Jo. "A Girl Was Allegedly Raped in the Metaverse. Is This the Beginning of a Dark New Future?" In: *The Guardian* (Jan. 2024). ISSN: 0261-3077. (Visited on 01/20/2024).

ScaDS_PubRel. *ScaDS.AI - Center for Scalable Data Analytics and Artificial Intelligence.* https://scads.ai/cracking-the-code-the-black-box-problem-of-ai/. (Visited on 01/20/2024).

Sean Michael Kerner. *Colonial Pipeline Hack Explained: Everything You Need to Know.* https://www.techtarget.com/whatis/feature/Colonial-Pipeline-hack-explained-Everything-you-need-to-know. (Visited on 01/20/2024).

Seven Ways Misinformation Spread during the 2016 Election. https://knightfoundation.org/articles/seven-ways-misinformation-spread-during-the-2016-election/. (Visited on 01/22/2024).

Shevchenko, Nataliya. *Threat Modeling: 12 Available Methods.* https://insights.sei.cmu.edu/blog/threat-modeling-12-available-methods/. Dec. 2018. (Visited on 01/20/2024).

Smart Home. https://www.statista.com/topics/2430/smart-homes/. (Visited on 01/20/2024).

Tanya Basu. *The Metaverse Has a Groping Problem Already.* https://www.technologyreview.com/2021/12/16/1042516/the-metaverse-has-a-groping-problem/. (Visited on 01/20/2024).

Temple-Raston, Dina. "A 'Worst Nightmare' Cyberattack: The Untold Story Of The SolarWinds Hack". In: *NPR* (Apr. 2021). (Visited on 01/20/2024).

Thomas Brewster. *Fraudsters Cloned Company Director's Voice In $35 Million Heist, Police Find.* https://www.forbes.com/sites/thomasbrewster/2021/10/14/huge-bank-fraud-uses-deep-fake-voice-tech-to-steal-millions/. (Visited on 01/20/2024).

Threat Modeling Manifesto. https://www.threatmodelingmanifesto.org/.

Timothy Holman. *Paris: An Intelligence Failure or a Failure to Understand the Limits of Intelligence?* https://rusi.orghttps://rusi.org. Jan. 2024. (Visited on 01/29/2024).

Tims, Anna. "'Sim Swap' Gives Fraudsters Access-All-Areas via Your Mobile Phone". In: *The Guardian* (Sept. 2015). ISSN: 0261-3077. (Visited on 01/20/2024).

Triton Cyber Attack: Hackers Target the Safety Systems of Industrial Plants | SCOR. https://www.scor.com/en/expert-views/triton-cyber-attack-hackers-target-safety-systems-industrial-plants. (Visited on 01/20/2024).

UK Exposes Sick Russian Troll Factory Plaguing Social Media with Kremlin Propaganda. https://www.gov.uk/government/news/uk-exposes-sick-russian-troll-factory-plaguing-social-media-with-kremlin-propaganda. (Visited on 01/20/2024).

Ware, Willis H. "Security Controls for Computer Systems (U): Report of Defense Science Board Task Force on Computer Security". In: ().

Wolfendale, Jessica. "My Avatar, My Self: Virtual Harm and Attachment". In: *Ethics and Information Technology* 9.2 (July 2007), pp. 111–119. ISSN: 1572-8439. DOI: 10.1007/s10676-006-9125-z. (Visited on 01/20/2024).

The Easter Egg

Congratulations, you found the secret extra article. Admittedly it's not incredibly secret and well-hidden, but it is a little extra.

Easter eggs are a piece of computing history, and while becoming much less common for various reasons aren't likely to go away anytime soon. The first known computing Easter egg was around 1967-1968, in a compiler for the PDP-6 (a compiler turns programming code into executable code). The command for the compiler was 'make', at which point it would compile the target code. If told to 'make love', it would instead reply 'not war?' before continuing.

Despite this the term wasn't used to describe hidden or unexpected features in programs until 1979, about one hidden in the Atari game Adventure.

But this is a security book, so why am I going on about Easter eggs?

A.1 The Point

Every single vulnerability and exploit in technical security is down to unanticipated behaviour. No exceptions.

I could even make a solid argument that the same applies to any and all forms of security. That doesn't mean perfection is possible, but currently many of our technological systems are built with the assumption that all users are perfect in their actions, and that none of them are malicious.

There is always ongoing debate about breaking encryption for government surveillance. Recently, with the Online Safety Bill, a lot of very powerful lobbyists (and the politicians they lobbied) made an attempt to either ban or backdoor E2EE.

It was clear they hadn't[1] considered the consequences, instead being focused entirely on their goal. That goal is, unarguably, a laudable one – to protect children from abuse and exploitation. One, possibly the main, reason the debate around the topic was muted was because people trying to talk about the unintended consequences were often hit with the think of the children argument.

So, lets think of the children while we talk about the consequences of applying a backdoor in all E2EE encryption.

---

[1] And still haven't from the most recent debates I've been involved in.

A.2 Where is E2EE used?

The simple answer to this is everywhere. From posting on social media to your internet banking, almost every communication in the modern world uses some form of E2EE. Messaging is a little more complex, because some platforms instead provide only encryption until it hits their infrastructure, then re-encrypt to send to the recipient. This allows them to monitor, for whatever reason, the information you're sending.

Let's say that an authority has a backdoor for all E2EE. There are a couple of ways this could theoretically be done[2]. E2EE can allow for a third key, or more, to be used, so we could have one government 'master' key which is added to all E2EE communications. A second way would be to add an additional key to all communications, but a unique one for each that can then be accessed from a vault of some kind.

Some hybrid options exist, and there were suggestions of changing the mathematics of encryption so that it was possible for only an authorised third party to break it. As a note, and an important one, mathematics does not work that way so we'll ignore that option. The hybrid options provide various different balances of the drawbacks with the two approaches above.

---

[2] I say theoretically because the practicalities make it impossible

A.3   One Key to Rule Them All

If we have one master key, anyone with access to that key can decrypt everything. And that does mean everything. Given the motivation then to steal or copy it, there is no realistic chance it could be kept appropriately secure. All it would take is one analyst with access to it selling it on to criminals or a hostile power and suddenly everything would be effectively unencrypted. Bank transactions, card purchases, private messages, absolutely everything.

A.4   Challenges of Scale

If instead we have a key for each communication, we hit a scale problem. The sheer number of communications, most of which will use ephemeral keys with lifespans of minutes before being regenerated, is mind-boggling. We are talking thousands of keys per person per day. Even the storage, despite them being relatively small, would become a problem and you would still need to index them to each communication. There would be less motivation for someone to break in, largely because the system would be unusable.

A.5   Everyone is a Suspect

The problem with all of the approaches proposed is they treat everyone as a suspect, and expose everyone to oversight and surveillance. We know that these powers have been abused in the past, and we can guarantee they will be again. If they apply to everyone, the scale of abuse will be much worse.

The approach does not even help with the goal. The sheer investigative and enforcement resource that would be needed to validate every single alert from this sort of system is orders of magnitude larger than available today, and that's without considering the massive resources that would be consumed just by setting up and running such a system.

Effectively all it would achieve is to remove privacy from everyone, consume a lot of time and money, and most likely as with every other case where magic technological solutions are proposed for a human problem end up doing nothing to further their goals. An investment in the human investigation and enforcement capabilities instead would go far further, and likely less costly.

List of Figures

Glossary

Address Resolution Protocol A protocol which translates hardware addresses, which once were fixed but are now often randomly generated for phones and similar devices to provide a level of privacy, to network addresses.. 15, 224

Advanced Persistent Threat Groups of threat actors of varying degrees of organisation which range from coordinated state-funded, to commercially-driven cyber crime gangs, to anarchic hacktivist groups.. 79

Advanced Research Projects Agency The US Department of Defense's special projects arm.. 231

Anderson report The second of two volumes following the Ware report, which broadly outlines almost every principle of computer security (or cyber if you must) we struggle to properly today, written and published in 1972.. xix

Artificial Intelligence A poorly-used term most of the time which describes any technology which can make decisions in place of a human – so this includes everything from the latest Large Language Models (LLM)s to flow charts and decision trees. 46, 48, 68, 118–123, 131, 133, 134, 161, 167, 179–181, 184, 193, 226

attack tree A threat models approach based on starting with the goal and drawing branches of dependencies and methods to achieve them on the part of the attacker.. 36

baiting Usually used to reinforce a pretext by causing a problem, so for example pulling out a network cable discreetly and then being the friendly IT support person popping by because they saw a fault.. 12

Border Gateway Protocol A protocol built on shared routing tables which, when working, allows large segments of the internet to talk to each other. When it isn't working those sections of the internet break.. 114

bot Software used to carry out automated actions, from sending e-mails to pretending to click on advertising links to participating in massive Distributed Denial of Service (DDoS) attacks.. 66, 222–224

botnet A group of bots coordinated by a Command and Control (C2) infrastructure to act in concert.. 54, 65, 223, 224, 230

California Consumer Privacy Act The first of the US state privacy acts, modelled after the General Data Protection Regulation (GDPR).. 171

canary token A variety of techniques to raise an alarm and potentially gather additional information when a file or system is accessed or retrieved.. 166

catfish Now a less-used term for social media accounts or dating profiles set up to appeal to people, whether for romance scams or other purposes. I have no idea why it isn't spelled catphishing to stay in line with the other excessive jargon words for different forms of deception.. 58

Chief Technical Officer Usually an executive with responsibility for use of technology within a company.. 160

Command and Control Any of a number of methods used to control a number of bots or agents, usually referring to malware and botnets.. 54, 55, 222

Critical National Infrastructure Infrastructure technology considered critical, although definitions vary. In the UK relevant legislation means it includes digital marketplaces along with other sectors that might be more expected such as water and power.. 94, 95, 173

Denial of Service Any attack technique which causes a service or capability to fail.. 224

digital forensics Why this doesn't commonly have the acronym DF I will never know, even though that's how it's abbreviated in Digital Forensics and Incident Response (DFIR). Digital forensics is the examination of digital evidence of various types, which can be anything from simple log files to electron microscope scanning of cooled memory chips. While a vast number of technical techniques and tools are involved the key thing to remember, as in any form of forensics, is to preserve the chain of evidence.. 223

Digital Forensics and Incident Response A hybrid discipline made up of digital forensics and Incident Response (IR), designed both to deal with the aftermath of an incident and gather legally admissible evidence for any follow-up activities.. 200, 223

dis- and misinformation A general term to refer to both dis-information and misinformation for simplicity.. 57, 141, 191, 230

disinformation Deliberately designed misinformation to further an agenda, or simply opportunistically to worsen a target's position.. 67, 164, 166, 193, 198, 223, 226

Distributed Denial of Service A Denial of Service (DoS) attack which works through volumetric means, often using a number of bot in a botnet to overwhelm the capacity of a system.. 66, 222

Domain Name System The underpinning of the internet and the cause of many sleepless nights for sysadmins over the decades. DNS is used to resolve website addresses, which are friendly and human-readable, into the network addresses they are bound to. Those network addresses are then resolved to Address Resolution Protocol addresses to route traffic. Despite its essential nature, DNS is not as vulnerable as certain people repeatedly claim.. 55

dumpster diving A time-honoured tradition among hackers of rooting through dumpsters, disposal units, rubbish bins, or similar where documentation may be thrown away. Still effective today in the right circumstances.. 4

Electric Vehicle A battery-powered vehicle which, despite what marketing and investor hype may claim, is absolutely not safely self-driving. Cars with a traditional combustion engine of any type, including alternative fuels, are generally referred to as Internal Combustion Engine (ICE).. 58, 225

end to end encryption A form of encryption which protects a message transferred between two endpoints using a unique pair of keys, making any decryption of the message by any point in its training challenging.. 212, 213

Fear, Uncertainty, and Doubt An acronym for a type of marketing or hype based on scaring people into obediently buying whatever they're being sold.. 131, 158

General Data Protection Regulation The privacy and data protection framework for the EU, usually seen as one of the first and most significant. Many later data protection regulations have followed much of the same structure.. 163, 170, 171, 222, 227, 228

grok A word used to describe the intuitive graspability of a system. Related to ease of use, but incorporating something of an automatic grasp of why the system is built that way.. 36

High Net Worth High Net Worth refers to individuals with liquid assets usually over $1 million.. 231

human intelligence Sourcing intelligence through human activity, typically anything from covert to overt surveillance and including placing or recruiting informants, gathering rumours, and interrogation among other methods.. 192

Incident Response Incident Response covers the incident lifecycle – from prevention (or attempted prevention at least) through detection, containment, eradication, and recovery back to a normal state. At least that's the theory, in practice it can be misused to describe any number of those stages.. 223

Industrial Control System Technological systems used to control industrial systems, from critical national infrastructure such as power generation to manufacturing and more.. 94, 95

Internal Combustion Engine A vehicle which uses an engine based on internal combustion, such as petrol, diesel, or hydrogen fuel. Used mainly to contrast with Electric Vehicle (EV). 58, 224

Large Language Model A particular approach to artificial intelligence which relies on compiling large amounts of data in a vector database, categorising it, and generating new coordinates in that space in response to prompts. This is a subset of Machine Learning (ML) called deep learning.. 45, 47, 131, 175, 178, 180, 221

legal basis Under the GDPR the legal bases are a number of different legal reasons under which personal data may be processed. Not having a legal basis for processing is a breach of the act.. 170

love bomb A term for a particularly nasty technique used in everything from Multi-Level Marketing (MLM) recruitment to radicalisation, and other abusive relationships of all types. It consists of overwhelming a target with compliments and support.. 59, 61

Machine Learning An approach to building Artificial Intelligence (AI) based on allowing machines to build models which recognise patterns, usually involving training data and reward/punishment approaches. The rewards and punishments are mathematical weightings, and do not involve beating computers with sticks.. 45, 46, 50, 179, 226

malware A portmanteau of and generic name for malicious software.. 54, 96, 228

man-in-the-middle A type of attack involving interception of communications – while talked about as if they are astoundingly common, the average attacker will struggle to gain access to infrastructure and these attacks more often occur where someone has set up a malicious wireless network.. xxi, 15–17, 22, 24

misinformation False information erroneously created and spread. Deliberately created and spread false information is disinformation.. 140, 166, 223

Multi-Level Marketing A business structure which is absolutely, definitely not a pyramid scheme (they absolutely, definitely are), usually using unethical recruitment techniques to draw people in and have them exploit friends, family, and acquaintances to pay back their fees for joining the scheme (other approaches exist).. 57, 60, 226

Open Source Intelligence An approach to intelligence gathering and research which relies on public sources and passive techniques which do not involve interaction with the subject.. v, 28–32, 166, 198

operational security Maintaining information security within an operation or project, originally military but adopted

pretexting To use a pretext for something, such as pretending to be technical support. Impersonation is also used at times by people who aren't quite as jargon-heavy.. 10, 12

principal Regular Circuit readers will not need this explained, but for the sake of others a principal in close protection and related professions is the person under protection.. 29, 141, 166

Privacy and Electronic Communications Regulations Similar to GDPR, a regulatory framework created by the EU and implemented in all member states, with the major impact on individuals being some degree of protection against unwanted electronic marketing.. 171

ransomware A form of malware which works by encrypting data and charging a ransom to decrypt it.. 87–91

reputation management A set of techniques and tools designed to ensure that a reputation is positive, variously by promoting actions that cast the subject in a good light, or suppressing ones that could be negative. It is occasionally seen as an unethical field given some of the techniques, such as Strategic Lawsuit Against Public Participation (SLAPP)s. 163

Security Incident Event Management A system which collates logs and events together to present them in a single interface for analysts, often with some automated rules to alert on particular events or combinations of events.. 159, 229

Security Operations Centre Often used outside of the cyber security industry to describe a CCTV control centre, or similar control centre. Within cyber security it generally refers to an operations centre where much of the logging information for an organisation is forwarded to a Security Incident Event Management (SIEM) and monitored by analysts.. 110

shoulder surfing Standing behind someone and watching over their shoulder (usually surreptitiously) to capture credentials or sensitive information.. 9

SMishing Another pointless jargon term for phishing via text message instead of e-mail.. 9, 10, 71

spear phishing Just yet another variant on phishing, this could more usefully be called targeted phishing.. 10, 71, 231

Strategic Lawsuit Against Public Participation Lawsuits, or threats of lawsuits, designed to prevent criticism, publication of negative information, or discussion of certain issues. Used by companies, politicians, and sometimes even by individuals in the security and data protection industries. Some measures have been taken to reduce the impact, but there is a long way to go.. 228

STRIDE A threat models approach developed by Microsoft and centered around the acronym Spoofing, Tampering, Repudiation, Information disclosure, Denial of service, Elevation of privilege as a brainstorming tool.. 36, 38–40, 229

STRIPED Just STRIDE with Privacy added.. 39

tailgating A social engineering technique involving following someone through a security checkpoint or theoretically secure door casually and not being challenged.. 12

threat model An approach to security by design based on the revolutionary and under-practiced idea of considering what could go wrong before it does.. 35, 221, 227, 229

troll farm Troll farms are large buildings (or, in these days of remote and distributed work, organised groups of workers not sharing a building – though that approach seems unpopular with the type of government that uses troll farms) where people are employed to run social media profiles. They will create dis- and misinformation, amplify and re-post any form of content that suits their agenda, and provide the human touch to social media botnets. It sounds like a conspiracy theory, but they are well-documented. Also known as troll factories. In countries with low local wages you may also come across click farms, similar setups where people are paid to click on advertising links in huge quantities.. 67, 164

Venture Capital A form of private equity funding, providing funds to companies from startups to ones emerging on the market which are expected to have a high degree of growth to provide a return on investment.. 157, 160, 161

virtual private network A technology used to create a private, theoretically secure, connection between two endpoints. The idea is to allow secure communication across an untrusted network so long as the network from the endpoint onwards is trusted.. 193

vishing Yet another pointless jargon term, for phishing via a phone call. Once upon a time this would have just been phone fraud, but now we need a special term to market expensive solutions.. 9–11, 71

Ware report A foundational text in computer security by Willis Ware, produced on commission from Advanced Research Projects Agency (ARPA). xix, 221

watering hole attack A form of attack where instead of the attacker going to the target, they try to draw the target to them. An example would be an attack using phish offering a fake discount voucher to draw the target to visit a malicious website.. 10

whaling Yet another form of phishing, this one a narrow variation on spear phishing targeted at senior executives and other High Net Worth (HNW) individuals.. 10, 11, 71

Index

Credits

This book was helped along by a number of alpha and beta readers. So, special thanks to:

- John Amer

- Gerald Benischke

- Sarah Lascelles

- Kat Samperi

- Rowan Troy

- Lisa Ventura MBE

- Angela Vernon-Lawson CSyP

All provided excellent feedback, and this book would not be nearly what it is without their help and support.

About the Author

James Bore MSc CSyP is an independent cyber security consultant, speaker, and author with over two decades of experience in the domain. He has worked to secure national mobile networks, financial institutions, start-ups, and one of the largest attractions' companies in the world, among others. He holds a Master of Science degree in Cyber Security from Northumbria University and is a Chartered Security Professional (CSyP).

James is a regular contributor to the Circuit Magazine, a leading publication for the security industry, where he writes about various topics related to cyber security and humans. He is passionate about raising awareness and educating people on the importance of security as a discipline and a skillset, not just a technical domain. He also enjoys sharing his insights and experiences through public speaking, podcasts, and social media.

In his spare time, James likes to read, play board games, practice amateur butchery, and cook. He is based in the UK, but works with clients all over the world. You can find out more about him and his work on his Linktree at https://linktr.ee/coffeefueled.

BORES

secure with us

https://www.bores.com

Bores is a family business first established in 1988 and at the cutting edge of technology and security consultancy ever since.

Now in the second generation, run by James Bore, they provide managed security services, awareness training, and compliance implementation.

The Circuit

MAGAZINE

https://www.circuit-magazine.com

Most of these articles are available in previous issues of the
Circuit Magazine, and more will be coming in the future.
Alongside there are other excellent articles on everything from
close protection, to international security, and everything
security in between.

Subscribe today!

Security Blend Books

Security Blend Books is a boutique publisher for the security industry. Working with modern print on demand technology, we provide not only a route to market for new and upcoming security authors, but expert subject matter expertise on the field.

To find out more about us, other upcoming books, and how our submission and editorial process works, go to our website.

https://securityblendbooks.com